4.44

# CONTRACT ADMINISTRATION

## FOR THE BUILDING TEAM

### EIGHTH EDITION

*Other titles by the Aqua Group*

*Pre-Contract Practice for Architects and
Quantity Surveyors*

*Tenders and Contracts for Building*

# CONTRACT ADMINISTRATION

## FOR THE BUILDING TEAM

### EIGHTH EDITION

## THE AQUA GROUP

**With sketches by
Brian Bagnall**

**Blackwell
Science**

Blackwell Science Ltd
Editorial Offices:
Osney Mead, Oxford OX2 0EL
25 John Street, London WC1N 2BL
23 Ainslie Place, Edinburgh EH3 6AJ
238 Main Street, Cambridge,
 Massachusetts 02142, USA
54 University Street, Carlton,
 Victoria 3053, Australia

Other Editorial Offices:
Arnette Blackwell SA
 224, Boulevard Saint Germain
 75007 Paris, France

Blackwell Wissenschafts-Verlag GmbH
 Kurfürstendamm 57
 10707 Berlin, Germany

 Zehetnergasse 6
 A-1140 Wien

First edition published by
 Crosby Lockwood & Son Ltd 1965
Second edition 1972
Third edition published by Granada
 Publishing Ltd in Crosby Lockwood
 Staples 1975
Fourth edition published by Granada
 Publishing Ltd 1979
Fifth edition 1981
Reprinted 1983
Sixth edition published by Collins
 Professional and Technical Books 1986
Seventh edition published by BSP
 Professional Books 1990
Reprinted 1992, 1994
Eighth edition published 1996
by Blackwell Science Ltd

Set in 10/12.5 pt Century
by DP Photosetting, Aylesbury, Bucks
Printed and bound in Great Britain by
Hartnolls Ltd, Bodmin, Cornwall

DISTRIBUTORS

Marston Book Services Ltd
PO Box 87
Oxford OX2 0DT
(*Orders:* Tel: 01865 791155
    Fax: 01865 791927
    Telex: 837515)

USA
Blackwell Science, Inc.
238 Main Street
Cambridge, MA 02142
(*Orders:* Tel: 800 215-1000
     617 876-7000
     Fax: 617 492-5263)

Canada
Copp Clark, Ltd
2775 Matheson Blvd East
Mississauga, Ontario
Canada, L4W 4P7
(*Orders:* Tel: 800 263-4374
     905 238-6074)

Australia
Blackwell Science Pty Ltd
54 University Street
Carlton, Victoria 3053
(*Orders:* Tel: 03 9347-0300
     Fax: 03 9349 3016)

A catalogue record for this book is available from the British Library

ISBN 0–632–03847–0

Library of Congress
Cataloging-in-Publication Data
is available

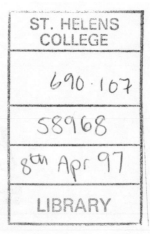

# Contents

"As Employer you might like to meet your builder."

bgb.

# Introduction

Although the membership of the Aqua Group has changed over the years – indeed, only two of the original members are now regularly involved – our objective has remained much the same as it always was: namely, to set out in clear, concise and practical terms, principles of good practice in the work of architects, quantity surveyors, contractors and all the other members of the building team.

Our belief in this concept has been reinforced by the fact that our three current titles, *Tenders and Contracts*, *Pre-Contract Practice* and *Contract Administration*, have long been established as standard reference works on good practice by experienced practising members of the building team. In addition, we are pleased to note that they are accepted as essential reading for students and are indeed recommended in many course reading lists.

Throughout the three books, we continue to assume the use of the JCT's Standard Form of Building Contract – a form that can be considered to be that tribunal's definitive statement on traditional contractual arrangements and one which is still used as the basis for teaching. We do, in addition, increasingly review alternative procedures, where these are in common usage. We also continue to assume that the contractor will generally be selected by competitive tendering. However, despite these assumptions, the principles expounded and the guidance given will invariably provide a basis for dealing with other forms of contract and other procurement procedures.

Our first book, *Tenders and Contracts*, analyses the basic concepts behind various methods of procurement and different types of contract. In examining the circumstances in which they may be appropriate, this book seeks to guide the reader in selecting the optimum way of providing the employer with value for money, within the constraints of the brief.

In *Pre-Contract Practice*, the important passage from inception through to the invitation of tenders is charted, with particular emphasis on developing the brief and translating it into tendering and construction documentation. This documentation must communicate the intentions of the designers to those responsible for pricing the tender documents and subsequently to those responsible for realising those intentions in material form – the completed building.

This third volume, *Contract Administration*, examines the period from the receipt of tenders to the settlement of the final account. It emphasises the importance of good communications in coordinating the efforts of the team to create that harmony so essential to the efficient running of a building project.

Such is the pace of change in building procedures, due to technological advances, and in building contracts, due to the involvement of the courts, that all three books are in a continuous state of review. However, we are confident that, while the tools available to do the job may progress towards greater efficiency and certainty, the strategic concepts do not change and still need to be enunciated regularly.

In preparing this, the eighth edition of *Contract Administration*, the Group has paid particular attention to those areas where practice has been affected by new legislation. We have updated the text to reflect the latest editions of JCT forms. We have also taken account of the increased scope of the team's involvement in assisting the employer to achieve his objective in the most effective way.

Chapter 1 has been expanded to include additional members in the building team, particularly the project manager and those responsible for implementing the Construction (Design and Management) Regulations (CDM), which came into effect on 31 March 1995. The concepts behind these regulations are examined at some length and the updated tabulation of contractual rights, duties and responsibilities has been moved to the end of the chapter, so as not to disrupt the text.

The importance of proper documentation, proper records and proper communications are emphasised, as one of the surest ways of avoiding disputes – or of protecting the employer's interests when disputes do arise.

In Chapter 2, the increasing tendency for contractors to qualify their tenders is discussed. This tendency is to be deprecated and should be unnecessary if the documents have been properly prepared and proper instructions have been given to the tenderers. We also continue to advise against the alteration of standard forms of contract, unless it is unavoidable. When such alterations are made, they should be carried out by specialists – so that the true intent is achieved and so that confusion and unintended consequences are avoided.

Reference to collateral warranties is now included, in the light of the increasing demand for their use by employers – for the benefit of third parties such as funding institutions and future tenants.

In Chapter 4, the section on site inspections has been expanded to reflect the increasing number of people who have a contractual or statutory right to visit the site and to clarify their various roles. The useful site check list is still included, although in slightly modified form.

In Chapters 5 and 7, the quantity surveyor's unilateral contractual responsibility in relation to the valuation of variations, and to the preparation of the final account, are emphasised. However, it is recognised that, in practical terms, the quantity surveyor will work with the contractor in expediting the production, and agreement, of the final account, thus avoiding potential disputes.

The final chapter, on insolvency, has been largely rewritten to give a more practical and structured approach to the resolution of the difficult situations which follow the insolvency of one party to the contract. The object is always to minimise any 'damage' and additional cost suffered by the innocent party, usually the employer.

Throughout the book, we have tried to keep the concepts and terminology simple, so that the reader quickly grasps the essence of the subject matter. In this context, we have used the terms *architect, engineer, quantity surveyor*, etc., without reference to gender, and to save repetition, have reverted to 'he' universally, where the sense allows. This, of course, is not to deny the existence of many able and eminent practitioners complying with the alternative specification! – indeed, we might echo the adage *'vive la différence'*. In like manner, in the contractual context, we have not adopted the now familiar, but long winded, format of *architect/contract administrator*. We trust our readers will support the sentiment behind these decisions.

The Group is grateful to the ICoW, JCT, the NJCC, the RIBA and the RICS for continuing to allow us to use their forms in various examples.

As always, the Group is indebted to Brian Bagnall for his delightful cartoons, which bring a smile to the lips, even when considering the most serious subject – our books would not be the same without them. The group would also like to acknowledge the contribution to our deliberations of Julia Burden, our Commissioning Editor.

The Aqua Group:

QUENTIN PICKARD BA RIBA (Chairman)

| | |
|---|---|
| BRIAN BAGNALL BArch (L'pool) | GEOFFREY POOLE FRIBA ACIArb |
| HELEN DALLAS BA DipArch RIBA | GEOFF QUAIFE ARICS |
| JOHN OAKES FRICS FCIArb | JOHN TOWNSEND FRICS ACIArb |
| RICHARD OAKES BSc(Hons) FRICS | JOHN WILLCOCK DipArch RIBA |
| ROBERT PEGG MBA ARICS ACIArb | JAMES WILLIAMS DA(Edin) RIBA |

'... vive la différence ...'

# Chapter 1
# The Building Team

Over the years, since the first edition of this book, the process of building has become more complicated. From interception to final completion, through site acquisition, design, contract and construction, each stage has become more time-consuming, and thus more expensive. The need to optimise this process is therefore of paramount importance, and the best base from which to achieve this, is proper and efficient team working.

It is therefore vital that all members of the building team are fully conversant, not only with their own role, but also with the roles of others and with the interrelationships at each stage of the project. Then each can play their part fully and effectively, contributing their particular expertise whenever required.

The make-up of any particular building team will depend on the scope and complexity of the works and the contractual arrangement selected. There are already many different methods of managing a project and, no doubt, others will be developed in the future. What follows is set in the context of JCT 80 and, although not exhaustive, will give an indication of the principles involved and the criteria by which other situations can be evaluated.

## Parties to a building contract and their supporting teams

The parties to a building contract are the employer and the contractor. Those appointed by these two will complete the 'building team' which can include:

the design team

- * employer
- project manager

*'The Building Team'*

- ● * planning supervisor
- ● * architect
- ● * quantity surveyor
- ●   structural engineer
- ●   building services engineers
- ● * nominated sub-contractor

the construction team

- ● * contractor (or principal contractor)
- ● * site agent (or foreman, described in the contract as the person-in-charge)
- ● * nominated sub-contractors
- ● * domestic sub-contractors

and in addition the employer may appoint a

- ● * clerk of works

It should be noted, however, that only those marked with an asterisk are mentioned in the contract.

This list is not exhaustive and to it could be added planners, landscape consultants, process engineers, programmers and the like. Equally, some roles may be combined and roles such as the project manager or planning supervisor may be fulfilled by individuals or firms from varying technical backgrounds.

## Rights, duties and responsibilities

JCT 80 is comprehensive on the subject of the rights, duties and responsibilities of the parties to the contract, their consultants or their representatives. As noted above, not all the members of the building team are mentioned in the contract, but those not mentioned will usually be given delegated responsibility from those who are mentioned. This delegation must be spelt out elsewhere in the correct documents – usually in the bills of quantities.

Whatever the number involved, each member of the team should be familiar with the contract as a whole and, in particular, with those clauses directly concerning their own work, so that the contract can run smoothly and efficiently. It should be noted that some of the duties are discretionary, some are mandatory and some even have statutory backing.

The rights, duties and liabilities of individual members of the building team are set out in a schedule at the end of this chapter, by reference to the private edition with quantities of JCT 80.

Other important considerations, relating to members of the building team but not necessarily codified in the contract conditions, are set out under specific headings below.

## Named consultants

While the architect and the quantity surveyor are referred to in the contract and are required to exercise specific responsibilities (many clauses include the phrase 'the architect shall'), they are not parties to the agreement. Should the contractor have a grievance, if they fail to carry out their duties, as defined in the agreement, the only contractual recourse is to seek redress from the employer.

## Un-named consultants with delegated powers

The project manager and the structural or other consulting engineers are not referred to in the contract, nor do they have any powers under the contract. Their position within the building team depends on the form of agreement they have with the employer or the architect. Where they have delegated duties, perhaps for design or site inspection, they should be named in the bills of quantities and the extent of the delegation should be identified, so that they have contractual recognition. Since they have no powers under the contract, if they need to issue instructions, this must be done through the architect.

As with the named consultants, if the contractor has any grievance against un-named consultants, redress can only be sought from the employer.

## The project manager

The project manager, who may be considered as the employer's representative, is likely to be appointed at the outset by the employer. The project manager, who is directly accountable to the employer, will be responsible for the successful outcome of the project in its broadest sense, from inception to completion. This will involve the programming, monitoring and management of the overall project, and advice to the client on all matters relating to the project, including the appointment of the architect, quantity surveyor and other consultants. However, not being mentioned in the contract, the project manager's position in respect of the contract works, which will be only a part of his overall duties and responsibilities, must be clearly determined and described in

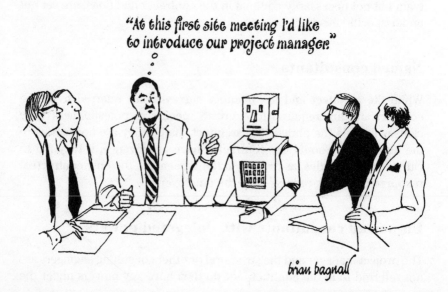

"At this first site meeting I'd like to introduce our project manager."

brian bagnall

the bills of quantities. Contractually, powers delegated should not exceed those set out in the contract in respect of the employer.

To what extent the role of the project manager, where appointed, will include, or be affected by, that of the planning supervisor, remains to be seen, but clearly the financial implications of employing these consultants may well influence the employer's approach to the project from the outset.

## The planning supervisor

The Construction (Design and Management) Regulations 1994 (CDM94), which came into effect on the 31 March 1995, introduced a new party to the building team, the 'planning supervisor'. Appointed by the client, the planning supervisor has overall responsibility, under the regulations, for co-ordinating the health and safety aspects of the design and planning phase of the project, and for the early stages of the health and safety plan and health and safety file, which are requirements under these regulations. The planning supervisor is specifically referred to in the contract as appointed by the employer, pursuant to regulation 6(5) of the CDM regulations.

In the contract, under Article 6.1, the planning supervisor is stated to be the architect, unless some other individual is identified.

## The principal contractor

In the contract, under Article 6.2, the principal contractor is stated to be the contractor or such other contractor as the employer shall appoint as the principal contractor pursuant to regulation 6(5) of the CDM Regulations.

The principal contractor will be a contractor, being a person or firm managing a construction contract for which they have been appointed by the employer. The employer must be reasonably satisfied that such principal contractor is competent and can allocate sufficient resources to ensure compliance with the regulations. The definition of a principal contractor in the CDM regulations allows for considerable flexibility, but it can be assumed that for the majority of construction contracts carried out under JCT 80 the main contractor will be appointed as the principal contractor. JCT Practice Note 27 appears to confirm this to be the intention of the JCT.

Where problems could arise, would be, for example, where pre-contract works, such as demolition or piling, are required before the appointment of the main contractor, or where a fitting-out contractor takes over after completion by the main contractor. In these circumstances, since there must always be a principal contractor in post while work is in progress on site, the employer will probably appoint a succession of principal contractors and care will need to be taken in passing responsibility from one to another.

The principal contractor is required to take over and develop the health and safety plan, to co-ordinate and activities of all contractors so that they comply with health and safety law and to provide appropriate information to the planning supervisor for incorporation in the health and safety file.

## Nominated sub-contractors

Nominated sub-contractors feature as members of the design team and as members of the construction team. This is because, as well as carrying out work on site, they are often involved in the design and planning of specialist works in advance of the appointment of the main contractor. Because of this special relationship, nominated sub-contractors acquire rights under JCT 80 which are not afforded to domestic sub-contractors, but by the same token, they have responsibilities which must be discharged in close liaison with the rest of the building team.

In JCT 80, nominated sub-contractors' special rights to payment are safeguarded by imposing responsibilities on the quantity surveyor, the architect and the contractor.

## The clerk of works

The clerk of works is normally appointed by the employer to act, under the direction of the architect, solely as an inspector of the works. Traditionally, he is likely to be an experienced tradesman, perhaps a carpenter and joiner or bricklayer. However, with today's highly complex and high-tech buildings, the architect, who will normally recommend the appointment, may need a technically experienced or qualified person and here the Institute of Clerks of Works will be able to assist. The clerk of works should be ready to take up his duties some time before the date for possession (how soon will depend on the size and complexity of the project), and will usually be resident on site for the duration of the contract, or will visit on a regular basis.

## Statutory requirements

As mentioned above, under the contract, the parties and the various consultants are charged with many duties, some of which are discretionary (the architect may...) and some of which are mandatory (the contractor shall...). In addition, provision is made for compliance with duties and responsibilities imposed by legislation (statutory matters). These may change from time to time or may be augmented by the addition of regulations issued by government under 'enabling Acts'. Constant vigilance is therefore required in keeping up-to-date.

Under the contract, 'Statutory obligations, notices, fees and charges' are covered generally in clause 6. Specific reference is made to VAT in clause 15 and to the statutory tax deduction scheme in clause 31. In March 1995, Amendment 14 to JCT 80 incorporated, via clause 6A, the Construction (Design and Management) Regulations 1994, commonly referred to as the CDM Regulations. These regulations are due to have a significant effect on all members of the building team – see below.

## CDM Regulations

The CDM Regulations place new duties on employers (termed clients under the regulations), planning supervisors, designers and contractors, to plan, co-ordinate and manage health and safety throughout all stages of a construction project.

No new laws on health and safety are introduced by the regulations, they merely codify the administrative machinery to provide an audit trail to demonstrate accountability for the management of health and safety throughout the project – very much in the same way that quality assurance under BS 5750 codifies general management procedures, except that contravention of the CDM Regulations can carry a criminal penalty.

The regulations will apply to most projects, but there are a few exceptions – notably, they will not apply:

- to construction work when the local authority is the enforcing authority for health and safety purposes
- to work which is not expected to last more than 30 days and involves four or less people on site at any one time
- to work for a domestic client, unless a developer is involved – apart from Regulation 7 (site notification) and Regulation 13 (designer duties).

Under the overall guidance of the planning supervisor, it is the designer who will be the responsible person in the early stages of a

8   *Contract Administration for the Building Team*

project, and while such duties will obtain throughout the design and construction periods, it is in the pre-contract stages that the majority of the designer's responsibilities under the regulations should be discharged. It is the designer's duty to tell the client initially what the client's duties are. The designer should design to reduce, if not avoid, as far as reasonably practicable, risks to health and safety both during construction and during subsequent occupation, and ensure that the design documentation includes adequate information on health and safety. Such information should be included on drawings or in specifications, and should be passed to the planning supervisor for inclusion in the health and safety plan.

Until the interpretation and respective responsibilities of the various parties under the regulations becomes clearer, the designer should exercise care in taking on the overall responsibilities of the planning supervisor. The indications are that these are not likely to be easily carried out by one particular discipline, and conflicts of interest could well arise. Article 6.1 of JCT 80 appears to recognise this by making the choice of the architect as the planning supervisor optional. Similarly the designer should not get involved with the principal contractor's duties; the regulations are quite clear that the management of safety during construction is the principal contractor's responsibility.

## Avoiding disputes

In *Tenders and Contracts for Building*, reference is made to the importance of formal procedures and proper documentation in the efficient and smooth running of building contracts – clarity and transparency are perhaps the key words.

This is particularly relevant in the avoidance of disputes. As projects become more complex, costs increase, margins tighten and employers demand greater quality and financial control, the margin between success and failure narrows, and there is less flexibility to absorb the 'swings and roundabouts' which have ever been a feature of the construction industry. The likelihood of disputes arising is, therefore, increased.

The building team must be vigilant, and ensure that the procedures employed, and the working relationships built up, produce an environment of co-operation, rather than discord.

Just as the building team strives for perfection in its working relationships, so the JCT strives to refine its contracts to take account of current practice and the latest decisions in the courts or arbitration proceedings. Hence the frequent revisions to the standard forms, which aim to achieve clarity and certainty in formalising the relationships between the parties.

In the event that disputes do arise, despite the best endeavours of the parties, the importance of proper documentation and compliance with formal procedures cannot be overstressed. If dispute proceedings become inevitable, it should be some comfort to know that proper documentation will be an asset rather than a liability.

On the assumption that the contract documents are complete, that tenders are reasonable and reflect current rates, and that information is available when required and not subject to late changes on the part of the employer or the architect, the origin of contractual disputes is seldom found to be in the dishonesty or incompetence of any party, but rather in the failure of one member of the team to convey information clearly to another. Unfortunately, what is not clear in the mind of the architect, for example, will be 'misty' to the quantity surveyor and 'foggy' to the contractor. This can lead to all sorts of problems!

Conveying intentions and instructions clearly is vital in the successful management of a contract, and all too often it is the breakdown, or failure, of communications that leads to dispute.

## Communications

Many books, conferences and papers have been devoted to the subject of 'communications' and it is a matter that cannot be dealt with exhaustively here. However, set out below are certain 'golden' rules to be observed by all members of the team in their dealings with each other:

- Do not tamper with the standard clauses of the building contract – if, however, the client insists upon it, employ a specialist.

- Ensure that the contract is executed prior to any start on site.

- Ensure that all team members have certified copies of the contract documents.

- Use realistic figures in the appendix to the contract.

- Where alternatives exist, or where entries have to be made in the appendix, identify such information in the bills of quantities at tender stage.

- Issue all instructions to the contractor through the architect.

- Issue all instructions to sub-contractors or suppliers through the contractor.

- Use standard forms or formats for all routine matters such as instructions, site reports, minutes of meetings, valuations and certificates, preferably with sequential numbering.

- Ensure that all verbal instructions are confirmed in writing as soon as possible after the event.

- Ensure that everybody is kept informed, not just those who have to act.

- Be precise and unambiguous.

- Act promptly.

Examples of suggested standard layouts for the more important communications passing between members of the building team are given in later chapters.

## SCHEDULE OF IMPORTANT RIGHTS, DUTIES AND LIABILITIES AS THEY CONCERN INDIVIDUAL MEMBERS OF THE BUILDING TEAM

(Based on the JCT Standard Form of Building Contract (JCT 80) Private Edition, with Quantities)

### The employer

#### Articles

Name and address
Appointment of architect
Appointment of quantity surveyor
Status under statutory tax deduction scheme
Rights, liabilities and procedure in respect of arbitration
Appointment of planning supervisor
Appointment of principal contractor

#### Clause

4.1   Right to employ others if contractor does not comply with instructions
5.1   Custody of contract documents (Local Authority edition only)
5.7   Duties in relation to confidential nature of contract documents
5.9   Rights to receive 'as built' drawings

# The contractor

26.1 Right to recover loss and expense incurred by matters materially affecting progress of the work

Duties as to notification and information to be provided

26.4 Duties in respect of claims by nominated sub-contractors

27 Rights and duties in event of determination of contractor's employment by employer

28 Right to determine contractor's employment in the event of employer's defaults and grounds for determination

29 Rights and duties in respect of work on site not forming part of the contract

30.1 Right to payment under interim certificate within 14 days

30.3 Duties in respect of off-site materials included in interim certificates

30.5 Right to require employer to place retention money in separate bank account (Private edition only)

30.6 Duty to provide documents necessary for adjusting the contract sum

Right to receive copy of computation of adjusted contract sum

31 Duties in connection with statutory tax deduction scheme (where employer is a 'contractor')

34 Duties consequent on the finding of antiquities

35.2 Right to tender for nominated sub-contract work

35.4–35.9 Procedure and duties regarding nomination of sub-contractors

35.5 Right to object to nominated sub-contractor

35.13 Duty to discharge interim payments to nominated sub-contractors

Duty to provide proof of payment to nominated sub-contractors

35.24 Duties in connection with re-nomination after default or determination by nominated sub-contractor

Duty to obtain instruction before determination of nominated sub-contractor's employment

38–40 The rights and duties in relation to choices available to employer for dealing with fluctuations

36.2–36.5 Procedure and duties regarding nomination of suppliers

41 Rights and duties in respect of arbitration

42 Procedure and duties regarding performance specified work

**Supplemental provisions**

Rights and duties in regard to VAT

# The architect

**Articles**

Name and address
Duties and procedure in respect of arbitration

**Clause**

| | |
|---|---|
| 2.3 | Duty as to discrepancies between documents |
| 3 | Duty to include ascertained amounts in interim certificates |
| 4.2 | Duty to justify instructions |
| 4.3 | Duty to issue instructions in writing |
| | Procedure in connection with verbal instructions |
| 5.1 | Custody of contract documents (Private edition only) |
| 5.2–5.4 | Duties concerning furnishing copies of drawings and documents |
| 5.6 | Right to require return of drawings on completion |
| 5.7 | Duties in relation to confidential nature of contract documents |
| 5.8 | Procedure for issue of architect's certificates |
| 6.1 | Duty to issue instructions in connection with statutory requirements |
| 6A.4 | Duty in connection with the health and safety file required by the CDM Regulations where the architect is the planning supervisor |
| 7 | Duties as to setting out |
| 8.2 | Right to require proof of standards of materials |
| 8.3 | Rights as to opening up of suspect work |
| 8.4 | Rights as to removal of faulty work |
| 8.5 | Right to order exclusion of persons from the works |
| 11 | Right to access to job and workshops |
| 13.2 | Right to issue instructions requiring variations |
| 13.3 | Duty to issue instructions regarding provisional sums |
| 13A.3 | Duty to confirm in writing acceptance of a 13A quotation ('a confirmed acceptance') |
| 16.1 | Rights regarding removal of unfixed goods |
| 17.1 | Duty to issue certificate of practical completion |
| 17.2–17.5 | Duties as to defects |
| 18.1 | Duties concerning partial possession by the employer |
| 19.2–19.3 | Duties and procedure as to sub-letting by contractor |
| 21.1 | Right to require evidence of insurance by contractor |

| | |
|---|---|
| 21.2 | Duties in regard to insurances against damage to property (other than the works) where not caused by contractor's negligence etc |
| 22 | Duties and rights in connection with insurance of the works |
| 23.2 | Rights regarding postponement of work |
| 24.1 | Duties to issue certificates in event of non-completion |
| 25.3 | Duty to grant extensions of time and to fix new completion date: information to be given |
| 26.1 | Duty to ascertain loss and expense incurred by contractor |
| 26.3 | Duty to give details of extension of time granted |
| 26.4 | Similar duties in respect of nominated sub-contractors |
| 27.1 | Procedure for issuing notices in the event of default by contractor |
| 27.2 | Right to give notice in respect of default by contractor |
| 27.4 | Duties in event of determination of contractor's employment by employer |
| 30.1 | Duty to issue interim certificates |
| 30.3 | Discretion to include off-site materials in interim certificate |
| 30.5 | Duty to prepare and issue statements of retention in respect of each interim certificate |
| 30.6 | Duty to inform contractor and sub-contractors of final valuations of work of nominated sub-contractors |
| 30.7 | Duty to issue interim certificate including all final amounts due to nominated sub-contractors |
| 30.8 | Duty to issue final certificate and to inform each nominated sub-contractor of date of issue |
| 34 | Duties relating to antiquities found on site |
| 35.4–<br>35.9 } | Procedure and duties regarding nomination of sub-contractors |
| 35.13 | Duties regarding interim payments to nominated sub-contractors |
| 35.14 | Duty to operate provisions of sub-contract in dealing with applications for extension of time |
| 35.15 | Duty to certify if nominated sub-contractor fails to complete in time |
| 35.16 | Duty to certify practical completion by nominated sub-contractor |
| 35.17 }<br>35.18 | Duties regarding early final payment to nominated sub-contractor |
| 35.24 | Duty to re-nominate if nominated sub-contractor defaults or determines sub-contractor's employment |
| 35.25 | Duties in connection with determination of nominated sub-contractor's employment |

# Sectional completion supplement

Where this supplement is incorporated in the contract to provide sectional completion by phases, the duties and rights of the employer, contractor, architect and quantity surveyor are clearly defined.

# Chapter 2
# Placing the Contract

The placing of the contract is a relatively simple routine matter, but the events which immediately precede it, and those which follow immediately afterwards, are of great importance.

## Reporting and tenders

The receipt and examination of tenders is discussed in *Pre-Contract Practice* in the context of the Code of Procedure for Single Stage Selective Tendering, published by the National Joint Consultative Committee for Building (NJCC). Under this Code, good practice requires that contractors submit unconditional or unqualified tenders, in accordance with the instructions issued to them. Such instructions usually require the submission of a single tender figure, but sometimes alternative figures are requested, based on different criteria – for example, based on different contract periods.

Qualified tenders are at risk of being rejected, on the basis that the tenderer has had ample opportunity of clarifying queries during the tender period and, not having done so, displays some ulterior motive. In such circumstances, what might have been a bona fide tender could be disallowed.

Despite instructions to the contrary, some contractors still submit qualified or alternative tenders incorporating specific conditions – for example, by employing an alternative form of construction, or specification, the contractor may be able to offer a cost saving or reduction in the contract period. Such submissions are contrary to good practice and should not be encouraged – they tend to be disruptive and to take extra time to resolve when reporting on tenders in a tight timescale.

The simple logic behind this principle, is that the alternative is almost certain to have been considered and rejected by the design team before tenders were sent out and that, if all tenderers had the opportunity of pricing the same alternative, the ranking of the tenderers would be unlikely to change.

If, contrary to all this advice, a contractor's alternative proposal is considered to have merit by the design team, suitable comments can be incorporated in the tender report.

In normal circumstances, the tenders will be checked, the quantity surveyor will examine the bills of quantities of the lowest tenderer and the design team will submit a report to the employer, with recommendations for action. The employer should be encouraged to make an early decision on the report, for it is of the utmost importance to a contractor to know quickly whether or not his tender has been successful.

If no serious errors have been found in the bills of the lowest tenderer, the design team's report will normally recommend acceptance of that tender. Where the widely accepted practice of single stage selective tendering has been followed, acceptance of other than the lowest tender should be considered only in the most exceptional circumstances. However, if the tender instructions have indicated that price will not be the only criterion by which offers are evaluated, in making their recommendation, the team will take into account other factors which may be more important than price, such as the contractor's approach to specific problems, his method of working, or his programme. Careful consideration may conclude that the lowest tender is not necessarily the best value for money and the team will recommend accordingly.

## Notification of tender result

It is usual for the employer to accept the team's recommendation and, as soon as he does so, all contractors who tendered should be notified and sent a list of the tenders received. If priced bills of quantities have been submitted at the same time as the tenders, these should be returned to the unsuccessful contractors unopened.

## Errors in bills of quantities

In the course of his examination of the priced bills of quantities, the quantity surveyor may have found some errors in pricing or arithmetic, or perhaps in both. All errors should be reported to the contractor and, usually, the method of resolving them will be agreed before the report is submitted to the employer.

The invitation to tender should state which of the alternatives, under section 6 of the NJCC Code of Procedure for Single Stage Selective Tendering 1989, is to apply. Under alternative 1, the contractor has to withdraw, if he is not prepared to stand by his tender, while under alternative 2, he has the opportunity of correcting genuine errors.

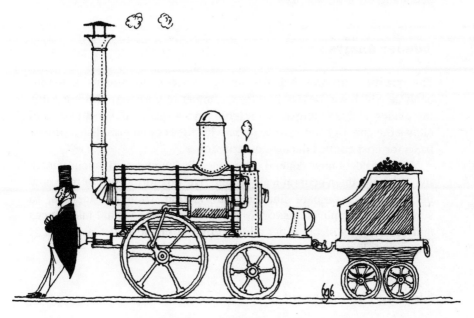

*'... agreed to stand by his tender ...'*

Whichever procedure applies, errors must be dealt with in the appropriate way, in order to put the bills of quantities right for their use as a contract document.

If alternative 1 applies and the contractor has agreed to stand by his tender, the errors should be put right, the arithmetic corrected and the summary amended as necessary. An adjustment should then be made at the end of the summary which will leave the final total of the bills equalling the original tender figure. This adjustment will be a lump sum equal to the net amount of the errors and will effectively be added to, or deducted from, the corrected total of the summary. Unless the error is very small and can be adjusted against a single lump sum in the priced bills (the figure for water for the works, or insurance, is often used), a note should be added in which the amount of the adjustment is expressed as a percentage of the total value of the general contractor's work (i.e. the total of the bills, less preliminary items, contingencies and prime cost and provisional sums). Any rates in the bills subsequently used for valuing the contractor's work in variations or interim certificates will then be adjusted by this percentage.

If alternative 2 applies, the errors will be corrected in the same way and the summary amended, but no adjustment will be made to restore the total to the original tender figure, the revised summary total becoming the contract sum.

## Tender analysis

The quantity surveyor will frequently be asked to prepare a tender analysis which can be required for a number of purposes. Traditionally, the tender analysis compares tendered prices against the cost plan, and allows the quantity surveyor to update his figures and establish a proper basis for cost control during construction.

Many projects now depend upon outside funding or financial incentives to make them economically viable. The quantity surveyor may now be required to prepare financial information, from his analysis, to support grant applications, or to take advantage of capital allowances and other forms of tax relief allowed by the Inland Revenue and H M Customs & Excise.

## Signing the contract

While the contract documents are being prepared, the architect should agree with the contractor the dates for possession of the site and completion of the work, if these have not already been decided. The architect should also make arrangements with the contractor for the initial site meeting and should ensure that the contractor has no valid objection to any nominated sub-contractor or supplier.

Once these matters have been dealt with, the contract documents should be signed. These normally consist of the Articles of Agreement, the drawings showing the extent and nature of the works and the bills of quantities, all of which should be initialled. Any post tender negotiation documentation should also be attached and initialled.

It must be borne in mind that, in addition to completing the Articles of Agreement at the front of the JCT 80 form of contract, it is also necessary to make a number of deletions or amendments in the text of the conditions. These are as follows:

● Clause **5.3.1.2** Under this clause, the contractor is required to provide the architect with copies of his master programme for the execution of the works and subsequently to amend it in the event of an extension of time being granted. If this is not required, this clause should be deleted and, in the following clause (5.3.2), the words in parentheses, which refer to the master programme, should also be deleted.

- Clauses **22A**, **22B**, **22C** Two of these clauses must be deleted, according to whether the employer or the contractor is responsible for the insurance of the works. There is a reminder in a footnote to these clauses that it is sometimes not possible to obtain insurance in the precise terms required by the contract, in which case the contract must be amended accordingly.

- Clause **35.13.5.3.4** This clause will require amendment, as indicated in the footnote, in the rather unlikely event of the contractor being an individual or a company not incorporated under the Companies Acts.

- Clause **41.7** This clause needs amending if the parties do not want the law of the contract to be the law of England and/or do not want the Arbitration Acts 1950 and 1979 to apply to the resolution of disputes.

These deletions and amendments should have been notified to the contractors in the bills of quantities at the time of tendering and no other deletions or amendments should be made without the prior agreement of the contractor. All deletions and amendments must be initialled by the parties at the time the contract documents are signed.

It is also necessary to complete the appendix to the conditions of contract, and here again the information to be inserted should have been stated in the bills of quantities. In the event of any item in the appendix having been left for decision until after tenders have been submitted, the matter must be agreed with the contractor before the contract documents are completed.

There are three supplements to JCT 80 which must also be dealt with at this stage. They are:

- supplemental provisions (the VAT Agreement)
- sectional completion supplement and practice note 1
- contractor's designed portion supplement

The first of these is the supplemental provisions dealing with VAT, which is bound in with JCT 80. The appropriate section in the appendix must be completed.

The second supplement is appropriate in dealing with sectional completion of the works, which will be discussed later.

The third is printed as a separate document and is used if the employer requires the contractor to undertake the design of a portion of the works and operates in a similar fashion to the JCT Standard Form of Building Contract with Contractor's Design (JCT 81). The contractor is required to submit his design proposals with his tender, and any default (which is not uncommon) must be rectified. JCT Practice Note CD/2, available as a separate document, gives further guidance on the use of this supplement.

Amendment 12 to JCT 80 introduced Part 5, Performance Specified Work (clause 42), which closely mirrors the operation of the contractor's designed portion supplement. However, in this alternative approach to the introduction of an element of contractor's design, the contractor is not required to put forward any design proposals for performance specified work until the contract has been let. Performance specified works should normally be limited to fairly straightforward packages, such as pre-cast concrete floor units, trussed rafters, or simple services installations. JCT Practice Note 25 deals with performance specified work in detail.

## Sectional completion

Returning to the sectional completion supplement, which is printed as a separate document, it is important to distinguish between this and clause 18 of JCT 80 – Partial Possession by Employer. The sectional completion supplement enables JCT 80 to be adapted for use where the works are to be completed by phased sections. It can only be used where tenderers are notified that the employer requires the works to be carried out in phased sections, with the employer taking possession on the practical completion of each section. If the work has not been divided into sections in the tender documents, the supplement cannot be used. In such cases, if the employer wishes to take possession of parts of the work during the course of the contract, the provisions of clause 18 will apply, but it must be noted that, under that clause, prior consent of the contractor must be obtained.

If provision has been made for including the sectional completion supplement in the contract documents, the first page of the Articles of Agreement in the form must be amended or replaced by the equivalent page in the supplement. In addition, the appendix in JCT 80 must be deleted and replaced by the supplement appendix, which must be completed in accordance with the notes in the supplement (Practice Note No. 1 is bound into the document).

## Executing the contract as a deed

Prior to preparation of the contract documents, both parties should be consulted as to whether or not the contract is to be executed as a deed. Many parties to contracts prefer to have them executed as a deed in order to obtain a 12 year period in which to commence actions for breaches of contract, instead of the six year period applicable to contracts under hand. These limitation periods are dictated by the Limitation Act 1980. In the event of one party executing the contract as a deed and the other signing it, the 12 year limitation appears to apply only against the party who executed it as a deed. By virtue of the Finance Act 1985 stamp duty no longer applies to contracts executed as a deed.

The Limitation Act 1980 also deals with limitations for actions in tort, and the Latent Damage Act 1986 amends the 1980 Act by setting time limits for negligence actions in respect of latent damage not involving personal injuries. Although not entirely clear from the 1986 Act itself, there are grounds to support the view that it applies to actions in tort and not in contract. The time-limits for commencing actions of the type dealt with by the 1986 Act are:

- six years from the actual occurrence of damage
- if later, three years from when the damage could reasonably have been discovered, as set out in the Act

Both of these, however, are over-ridden by a 15 year long-stop period running from the occurrence of the negligent act or omission causing the damage. These time limitations should be treated with some caution, as recent court judgements have introduced a degree of uncertainty.

Clause 5 of the conditions of contract requires that the contract documents be held by the architect or quantity surveyor so as to be available at all reasonable times for inspection by the employer or the contractor. However, as previously noted, it could be very helpful for copies of appropriate parts to be made available to those members of the building team who require them.

## Performance bonds or parent company guarantees

A performance bond is a three party agreement between the employer, the contractor and a surety, who agrees to pay a sum of money to the employer, in the event of default by the contractor. The effectiveness of performance bonds is being called into question by recent court decisions. For example, bonds are frequently thought to provide protection to the employer against the insolvency of the contractor, but the courts have held that insolvency of the contractor does not amount to a breach of contract, as there are procedures within the contract to deal with this event – consequently the bond could not be called in. (Adjustment to the wording of the bond may overcome this problem.)

The NJCC has issued Guidance Note 2: Performance Bonds, which briefly explains the subject and gives an example, which is reproduced at the end of this chapter as *Example 1*.

A parent company guarantee serves a similar purpose to a bond, in that it attempts to protect one party to a contract from the effects of default by the other party, when the latter is a subsidiary company having limited assets. An example is given at the end of this chapter in *Example 2*.

It is important, where a bond or parent company guarantee is to be provided, that the details should be examined by the design team, and possibly the employer's legal advisors, before signing the contract, to ensure that it fully covers the requirements set down in the tender documents. (See also Chapter 9, Insolvency.)

## Collateral warranties

The growth in the demand for collateral warranties has provoked considerable debate within the construction industry and consultants are now frequently required to enter into collateral warranties as a condition of their principal appointment.

A collateral warranty is an agreement which runs alongside (or collateral to) a primary or principal agreement (the appointment). They are commonly used to bind a consultant into contract for a specific period of time with a third party, frequently the funding institution, where no contract would otherwise exist. (See also chapter 1 of *Pre-Contract Practice*).

The warranty will usually be required to be backed up by professional indemnity (PI) insurance. Standard forms of warranty are now available which have been drawn up and agreed by the principal parties and are generally acceptable to PI insurers. Non standard forms should be treated with caution, and it would be prudent to obtain legal advice before the warranty is given.

NJCC Guidance Note 6: 'Collateral Warranties' is a very useful document which explains the reasons behind the increasing demand for Warranties, and describes appropriate terms and conditions.

# Issue of documents

In accordance with clause 5 of the conditions of contract, immediately after the contract has been executed, the architect must furnish the contractor with:

- one copy of the contract documents certified on behalf of the employer
- two further copies of the contract drawings
- two copies of the unpriced bills of quantities.

As soon as possible following the execution of the contract, the contractor is to be provided with:

- two copies of any descriptive schedules
- two copies of any other like document necessary for use in carrying out the works
- two copies of further drawings or details to enable the contractor to carry out the works.

Providing clause 5.3.1.2 has not been deleted, as soon as possible after the execution of the contract the contractor shall provide the architect with two copies of the master programme.

Other documents which may be necessary for use in carrying out the works from the outset, include the following:

- party wall agreements
- schedules of condition of adjoining properties
- conditional planning permissions
- building regulation approval, including notices to be served during the course of the contract
- estimates from statutory authorities for services
- procedures or estimates for works to be carried out by local authorities, such as pavement crossovers and sewer connections.

# Insurances

The insurances required by the contract are dealt with in clauses 21, 22, 22A, 22B, 22C and 22D.

Clause 21 deals with three aspects:

(1) It reminds the contractor to insure his employees (as must anyone who employs people) under the Employer's Liability (Compulsory Insurance) Act 1969.

(2) It requires the contractor to insure against injury to, or death of, persons (other than his employees etc) to the extent of the cover entered in the contract appendix.

(3) It provides the employer with the option to require the contractor to take out a joint names insurance policy (i.e. the contractor and employer are jointly insured) to cover property other than the works, against damage caused by the carrying out of the works, excepting injury or damage:
  (a) caused by the contractor's negligence etc.
  (b) caused by errors or omissions in the designing of the works
  (c) which is reasonably foreseeable
  (d) arising from war risks or excepted risks i.e. risks which insurance companies exclude from the cover they are prepared to provide
  (e) covered by any insurance under clause 22C.1 taken out by the employer – see later

The extent of any cover will be as provided for in the contract appendix.

Clause 22 deals with insurance against damage to the works. The cover is against all risks except for those given as exclusions in clause 22.2. Clause 22A provides for the contractor to take out a joint names policy where the works comprise a new building. Clause 22B is used instead, if the works comprise a new building and the employer has elected to take out the joint names insurance policy. Clause 22C is used instead of 22A or B, if the works are an extension to, or an alteration in, an existing building. In this case the employer is required to take out two joint names insurances policies:

● to insure the works against all risks as previously described
● to insure the existing building other than the works (referred to in clause 22C.1 as existing structures and contents) against damage caused by the specified perils defined in clause 1.3.

The amount of cover is the full cost of the re-instatement, including professional fees and contents, as appropriate.

In respect of the contractor's insurance under clause 22A, it is common for the contractor to maintain an annually renewable insurance policy, which provides cover of no less than that required by clause 22A. Clause 22A.3 sets out deemed-to-satisfy provisions in respect of such annual policies.

The JCT has recently issued amendment TC/94 to the contract which deals with insurance cover against acts of terrorism. All risks insurance under contract clauses 22A, 22B and 22C include a requirement for cover in respect of fire and explosion. This does not include cover in the event of loss caused by acts of terrorism. Following terrorist attacks on mainland Britain, and particularly the IRA bomb in the City of London, insurance against terrorism has become unavailable. In order to comply with the contract requirement to maintain cover, the government has agreed to act as 'reinsurers of last resort' to enable cover to be maintained. This insurance cover will only normally be granted as an extension to the general insurance cover required under the contract. TC/94 sets out the various amendments to the contract to allow appropriate cover to be maintained. On existing contracts, where terrorist cover is no longer available, a supplementary agreement will be required to introduce the relevant amendments. The JCT Guide to Terrorism Cover provides further very useful information and advice.

Clause 22D provides the employer with the option to require the contractor to insure against delays to completion of the works caused by any of the specified perils defined in clause 1.3. The amount of cover will be at the rate stated in the contract appendix for liquidated and ascertained damages for the period given in the appendix. This type of insurance is currently expensive and, if required, is usually in respect of delays of no more than ten weeks.

Construction insurance is an extremely complex subject. It is further complicated by the fact that the policies available in the market do not have standardised wording. Advising on insurances is therefore very difficult and, for all but extremely simple situations, the client should be recommended to consult an insurance broker, or similar professional advisor.

Once insurances are in place, it is important that they are maintained and renewed when necessary. It is recommended that the architect or quantity surveyor ascertains directly, or through the employer's broker, that insurance policies are maintained and renewal premiums paid when due – copies of premium receipts, or a broker's confirmatory letter, should be kept on file.

# EXAMPLE 1: PERFORMANCE BOND

### MODEL GUARANTEE BOND

---

**THIS GUARANTEE BOND** is made as a deed **BETWEEN** the following parties whose names and [registered office] addresses are set out in the Schedule to this Bond (the "Schedule"):

(1) The "Contractor" as principal
(2) The "Guarantor" as guarantor, and
(3) The "Employer"

**WHEREAS**

(1) By a contract (the "Contract") entered into or to be entered into between the Employer and the Contractor particulars of which are set out in the Schedule the Contractor has agreed with the Employer to execute works (the "Works") upon and subject to the terms and conditions therein set out.

(2) The Guarantor has agreed with the Employer at the request of the Contractor to guarantee the performance of the obligations of the Contractor under the Contract upon the terms and conditions of this Guarantee Bond subject to the limitation set out in clause 2.

**NOW THIS DEED WITNESSES** as follows:

1    The Guarantor guarantees to the Employer that in the event of a breach of the Contract by the Contractor the Guarantor shall subject to the provisions of this Guarantee Bond satisfy and discharge the damages sustained by the Employer as established and ascertained pursuant to and in accordance with the provisions of or by reference to the Contract and taking into account all sums due or to become due to the Contractor.

2    The maximum aggregate liability of the Guarantor and the Contractor under this Guarantee Bond shall not exceed the sum set out in the Schedule (the "Bond Amount") but subject to such limitation and to clause 4 the liability of the Guarantor shall be co-extensive with the liability of the Contractor under the Contract.

3    The Guarantor shall not be discharged or released by any alteration of any of the terms conditions and provisions of the Contract or in the extent or nature of the Works and no allowance of time by the Employer under or in respect of the Contract or the Works shall in any way release reduce or affect the liability of the Guarantor under this Guarantee Bond .

4    Whether or not this Guarantee Bond shall be returned to the Guarantor the obligations of the Guarantor under this Guarantee Bond shall be released and discharged absolutely upon Expiry (as defined in the Schedule) save in respect of any breach of the Contract which has occurred and in respect of which a claim in writing containing particulars of such breach has been made upon the Guarantor before Expiry.

5    The Contractor having requested the execution of this Guarantee Bond by the Guarantor undertakes with the Guarantor (without limitation of any other rights and remedies of the Employer or the Guarantor against the Contractor) to perform and discharge the obligations on its part set out in the Contract.

6    This Guarantee Bond and the benefit thereof shall not be assigned without the prior written consent of the Guarantor and the Contractor.

7    This Guarantee Bond shall be governed by and construed in accordance with the laws of *[England and Wales] [Scotland]* and only the courts of [England and Wales] [Scotland] shall have jurisdiction hereunder.

## EXAMPLE 1: PERFORMANCE BOND *cont.*

**THE SCHEDULE**

The Contractor
The Guarantor
The Employer
}
*Names and addresses* or *(registered office addresses)*

The Contract       *Date – Form – Works – Value*

The Bond Amount       £ _____ *(words . . . . . . . . . . . . . . .)*

Expiry       *Agreed criteria to be used*

**IN WITNESS**
whereof the Guarantor and the Contractor have executed and delivered this Guarantee
Bond as a Deed this _____ day of _____ 199 .

**EXECUTED AND DELIVERED**       **EXECUTED AND DELIVERED**

as a deed by _____       as a deed by _____

in the presence of:       in the presence of:

_____ Director       _____ Authorised Signatory

## EXAMPLE 2: PARENT COMPANY GUARANTEE

**THIS AGREEMENT** is made the       day of       One thousand nine hundred and ninety-

**BETWEEN:**

**(1)**

      [of] **OR** [whose registered office is at]

      ('the Guarantor'); and

**(2)**

      [of] **OR** [whose registered office is at]

      ('the Employer', which term shall include its successors and assigns).

**WHEREAS** by an Agreement ('the Contract') dated       199 and made between the Employer of the one part and [       ] ('the Contractor') of the other part, the Contractor undertook the construction of certain Works in accordance with the terms and conditions of the Contract.

**NOW THIS DEED WITNESSETH** that if the Contractor defaults in the discharge of any of the Contractor's obligations under or pursuant to the Contract, the Guarantor will indemnify the Employer against all loss and damage thereby caused to the Employer, and no alterations in the Contract, or in the Works, and no extension of time, forbearance or forgiveness, nor any act, matter or thing whatsoever except an express release by Deed by the Employer, shall in any way release the Guarantor from any liability hereunder.

**IN WITNESS** whereof the Guarantor has executed this Deed on the date first stated above.

# Chapter 3
# Progress and Site Meetings

## Initial site or briefing meeting

As soon as practicable after the contract has been placed, the building team should meet. Although this initial meeting may take place on the site, it will more probably take place in either the architect's or contractor's office.

The manner in which this first meeting is conducted will greatly influence the success of the project and succinct clear direction from the chair will be a strong inducement to a similar response from the others. Since at this stage the person having the most complete picture of the job is likely to be the architect, it seems logical that the architect should take the chair. The arrangements for this meeting should be discussed with the contractor beforehand, and it is suggested that representatives of the following should attend, as appropriate:

- employer
- project manager
- architect
- quantity surveyor
- structural consultant
- services consultant
- contractor
- principal nominated sub-contractors
- principal nominated suppliers
- clerk of works

It is suggested that the agenda for this meeting should include the following matters:

- introduction of those attending
- factors affecting the carrying out of the work
- programme

- sub-contracts and employer/sub-contractor agreements
- lines of communication
- insurances (see Chapter 2)
- procedure to be followed at subsequent meetings

## Introduction

The introduction of those attending needs no elaboration, though it is more than just a formality as it establishes an initial contact between individuals who must work together in harmony if the contract is to run smoothly.

## Factors affecting the carrying out of the works

These would normally be described fully in the contract documents, but may require emphasis and clarification at this initial meeting. The factors may include:

- access to site
- space availability
- restrictions, such as hours of work and noise
- building lines
- buried services
- site investigation
- protection of the works, unfixed materials, adjoining buildings, work people and the general public
- Health and Safety at Work Act 1974, with particular reference to the Construction (Design and Management) Regulations 1994, which require the appointment of a planning supervisor

## Programme

The contractor should attend the initial site meeting with a draft or outline programme for the work prepared in advance, the necessary basic information regarding delivery dates and construction times having been obtained from the principal nominated sub-contractors and suppliers. It is helpful to have this programme circulated to those attending before the meeting when it can then be properly considered and adjusted as necessary. Following this, the contractor can then prepare the final programme and circulate it to all concerned. An example of a simple bar chart programme is shown as *Example 3* at the end of this chapter.

The opportunity should be taken to stress the importance of adhering to dates once these have been agreed by the contractor and the nominated sub-contractors and suppliers. This applies also to dates agreed for the issue of architect's or consultant's drawings, where these have not already been prepared in the pre-contract stage. It is not uncommon for the contractor to indicate on his programme the latest dates by which he requires drawings, instructions for placing orders, schedules and other information from the architect, and the latter must indicate at this stage whether the proposed dates are reasonable.

The master programme is not a contract document and clause 5.3.2 makes it clear that nothing contained in the programme can impose any obligation on the contractor beyond the obligations imposed by the contract documents as such.

If clause 5.3.1.2 of JCT 80 has not been deleted, it provides that, as soon as possible after the execution of the contract, the contractor shall provide the architect with two copies of his master programme for the execution of the works. Experience has shown that a master programme is an essential tool of management, both for the contractor and for the architect. Only in exceptional circumstances would a master programme not be necessary, and there is much to be said for the tendering contractors being required to submit a draft master programme with their tenders. Some forms of contract have no requirement for the contractor to submit a programme, in which case the need for a programme should be stated in the bills of quantities or the specification.

In addition to providing the master programme at the start of the contract the contractor is also required to update it within 14 days of any decision by the architect which may create a new completion date for the contract. It should hardly be necessary to stress the need to keep the programme up-to-date.

The judgment in *Glenlion Construction Ltd* v *The Guinness Trust* decided that:

(1) The contractor is entitled to finish the works before the completion date stated in the contract appendix (as extended by the provisions of clause 25).
(2) The contractor is not entitled to expect the design team to provide design information so that he can finish the works before the completion date stated in the contract appendix (as extended by the provisions of clause 25).

At first sight these two statements might appear ambiguous; but further thought will clarify that they are not because it may not be feasible for the employer and design team to provide design information quicker than would be necessary to complete the works by the contract completion date. If they can and all parties agree, then the solution might be for the employer and contractor to amend the contract to provide a mutually agreed shorter period for completion of the works.

## Sub-contracts

Many problems on building contracts arise from delay in the issue of instructions by the architect regarding nominated sub-contractors and supplies and from difficulties in liaison between the contractor and sub-contractors. It is advisable at this initial meeting, therefore, to clarify the position regarding all work covered by provisional and prime cost sums in the bills of quantities.

It is essential for the smooth running of the contract that all nominations are made in adequate time for the work concerned to be phased into the contractor's programme without causing disruption. It should be made clear to the contractor that, once nominated, these sub-contractors and suppliers are his responsibility contractually. It is important that the employer, the design team and the contractor allow sufficient time to carry out their various duties regarding the complex and often time-consuming procedures for the selection and appointment of nominated sub-contractors and suppliers.

The importance of proper sub-contract documentation should be stressed. This subject is dealt with in more detail in *Pre-Contract Practice*.

In the past, when nominated sub-contractors or suppliers defaulted and they had been responsible for some aspects of the design of their work and/or for the selection of materials and other such matters, it was usually impossible for the employer to obtain redress as the contractor had no such responsibilities under the main contract. In order to overcome this problem it is now common to arrange direct agreements, such as that in *Example 4*, between the employer and nominated sub-contractors or suppliers.

## Lines of communication

It is important at the initial site meeting that the procedure regarding architect's instructions should be made clear to all concerned. The matter is covered by clause 4 of the conditions of contract and is dealt with in Chapter 5 of this book. Additional points which should be stressed at the initial site meeting will be found in the golden rules of communications in Chapter 1.

# Subsequent site meetings (progress meetings)

Formal site meetings should not be confused with the architect's site visits and the numerous meetings between the contractor and others, which may be necessary during the progress of the work. The frequency of site meetings will vary with the size and complexity of the contract, and according to the particular stage of the job and any difficulties encountered. It is unlikely that these meetings would be at intervals of more than four weeks, and especially in the early stages they may well be held more frequently.

When considering the procedure to be followed, the following points should be borne in mind:

## (1) Notices to attend:

The architect should notify the main contractor, the quantity surveyor and consultants of the dates and times of site meetings, asking them to attend if their presence is required. It should be the responsibility of the contractor to call all sub-contractors' and suppliers' representatives whom he or the architect would like to be present.

## (2) Agenda:

The content of the agenda should be agreed before each meeting by the architect and the contractor. A standard form of agenda is useful, as a model and as an *aide mémoire*. A typical agenda is shown in *Example 5*.

## (3) Minutes:

Minutes should be impartially drafted and given the correct emphasis; they should be concise and should record decisions reached and action required. *Example 6* shows typical site meeting minutes.

The requirements of the employer and/or architect as to the nature, frequency, and procedures for, site meetings should be given in the main contract tender documents. On many large contracts it may be more convenient to divide these meetings into two parts: the first, under the chairmanship of the contractor, will deal with the method of carrying out the work. It would be attended solely by the contractor's representatives and those of sub-contractors and suppliers. It will plan and organise the work on site. The site meeting itself will monitor progress and performance. It will be chaired by the architect and be attended by the design team, the contractor, and such sub-contractors as are requested to attend. The object of this meeting would be to give the contractor the opportunity to raise with the team questions on the drawings, specifications, schedules and instructions issued; to request additional information and to report progress.

Consideration of the value of people's time should be shown at all meetings; care should be taken not to call to meetings those whose presence is not really necessary. It should also be remembered that queries, not requiring discussion within the team, should be sorted out immediately on the telephone or in correspondence with the appropriate member of the team – rather than waiting until the site meeting and wasting many people's time.

It is important to realise that a badly run site meeting can be a serious waste of time to all, but if properly handled, it can be a great aid to the smooth running of the contract. Furthermore, there is little doubt that these periodic meetings keep people up to the mark and impart a sense of urgency which is difficult in day-to-day correspondence.

'... *a great aid to the smooth running of the contract* ...'

The contractor should not regard meetings as relieving him of the obligation to manage the job efficiently.

## Employer's meetings

It is worth mentioning here that the employer may also wish to hold separate meetings off-site with the design team, to discuss general progress. Such meetings can act as a useful channel of communication for confirming various points which are often not decided until some while after work has started on site – for instance, colour schemes. Other points which can be discussed at these meetings include the effects of any variations to the works, and anticipated final costs.

# EXAMPLE 3: CONSTRUCTION PROGRAMME

**CONSTRUCTION PROGRAMME FOR:** ST SYLVESTER'S CHURCH HALL RENOVATION

↓ CONTRACT COMPLETION DATE ↓

| MONTH | NOVEMBER | | | | DECEMBER | | | | JANUARY | | | | | FEBRUARY | | | | MARCH | | | | APRIL | | |
|---|---|---|---|---|---|---|---|---|---|---|---|---|---|---|---|---|---|---|---|---|---|---|---|---|
| WEEK COMMENCING | 6 | 13 | 20 | 27 | 4 | 11 | 18 | 25 | 1 | 8 | 15 | 22 | 29 | 5 | 12 | 19 | 26 | 4 | 11 | 18 | 25 | 1 | 8 | 15 |
| WEEK NUMBER | 1 | 2 | 3 | 4 | 5 | 6 | 7 | 8 | 9 | 10 | 11 | 12 | 13 | 14 | 15 | 16 | 17 | 18 | 19 | 20 | 21 | 22 | 23 | 24 |

**OPERATION**

(CHRISTMAS HOLIDAYS / EASTER HOLIDAYS marked vertically in the chart)

- setting up site
- stripping out & making good
- padstones for new steelwork
- sample area for stone clean / main clean
- drain survey / new drainage work
- inspect extg roof / overhaul extg roof
- install new boiler room & kitchen
- new floor construction
- carpentry & joinery works
- new steelwork
- stairs & balustrading
- heating & mechanical (1st & 2nd fix)
- electrical (1st & 2nd fix)
- decorations
- final clean & handover

## EXAMPLE 4 *(page 1 only of seven page document included)*

Agreement NSC/W

# JCT

## JCT Standard Form of Employer/Nominated Sub-Contractor Agreement

Agreement between a Sub-Contractor prior to being nominated for Sub-Contract Works in accordance with clauses 35·3 to 35·9 of the Standard Form of Building Contract (1980 Edition incorporating Amendments 1 to 9 and Amendment 10) and an Employer.

[a] Insert the same details as in NSC/T Part 1, pages 2 and 3.

[a]   Main Contract Works ('Works') and location:

[a]   Job reference:

[a]   Sub-Contract Works:

[b] This Agreement must be executed before the Architect/the Contract Administrator can nominate the Sub-Contractor.

[b]   ## This Agreement

made the                    day of                              19

between

of (or whose registered office is situated at)

(hereinafter called 'the Employer') and

of (or whose registered office is situated at)

(hereinafter called 'the Sub-Contractor')

## Whereas

First     the Sub-Contractor has submitted a tender on Tender NSC/T Part 2 (hereinafter called 'the Tender') on the terms and conditions in that Tender and in the Invitation to Tender NSC/T Part 1 to carry out works (as set out in the numbered tender documents enclosed therewith and referred to above and hereinafter called 'the Sub-Contract Works') as part of the Main Contract Works referred to above to be or being carried out on the terms and conditions relating thereto referred to in the Tender NSC/T Part 1 (hereinafter called 'the Main Contract'); and the Tender has been signed as 'approved' by or on behalf of the Employer;

Second    the Employer has appointed

to be the Architect/the Contract Administrator for the purposes of the Main Contract and this Agreement (hereinafter called 'the Architect/the Contract Administrator' which expression as used in this Agreement shall include his successors validly appointed under the Main Contract or otherwise if appointed before the Main Contract is operative);

Third     the Architect/the Contract Administrator on behalf of the Employer intends that after this Agreement has been executed and, if a Main Contract has not been entered into, after a Main Contract has been so entered into, to nominate the Sub-Contractor to carry out and complete the Sub-Contract Works on the terms and conditions of the Tender and the Invitation to Tender NSC/T Part 1;

*continued*

# EXAMPLE 5: TYPICAL SITE MEETING AGENDA

| | |
|---|---|
| Project: | Shops & Offices, Newbridge St, Borchester |
| Project ref: | 456 |
| AGENDA FOR SITE MEETING | |
| Date: | 7 October 1997 at 10:00am |

| | |
|---|---|
| 1.0 | Apologies |
| 2.0 | Minutes of last meeting |
| 3.0 | Contractor report |
| | General report |
| | Sub-contractors' meeting report |
| | Progress and causes of delays and claims arising |
| | Information received since last meeting |
| | Architect's Instructions required |
| 4.0 | Clerk of Works report |
| | Site matters |
| | Quality control |
| | Lost time |
| 5.0 | Design Consultant reports |
| | Architect |
| | Structural Engineer |
| | Services Engineer |
| 6.0 | Quantity Surveyor report |
| 7.0 | Contract completion date |
| | Assess likely delays on contract |
| | Review factors from previous meeting |
| | List factors for review at next meeting |
| | Record completion date (as revised) |
| 8.0 | Any other business |
| 9.0 | Date, time and place of future meeting |
| | Site Meetings |
| | Site Inspections |

**Distribution:**

| Copies | 2 | Client | 1 | Structural Engineer |
|---|---|---|---|---|
| | 3 | Main Contractor | 1 | Services Engineer |
| | 1 | Quantity Surveyor | 1 | Clerk of Works |
| | 1 | Planning Supervisor | 1 | File |

---

# EXAMPLE 6: TYPICAL SITE MEETING MINUTES

| Project title | Shops & Offices<br>Newbridge St.,<br>Borchester | Reed & Seymore<br>Architects<br>12 The Broadway<br>Borchester<br>BC4 2NW |
|---|---|---|
| Project ref: | 456 | |
| Meeting title | SITE MEETING NO 4 | |

| | |
|---|---|
| Date | 7 October 1997 |
| Location | Site |

Those present:

| | |
|---|---|
| S Gilbert | Client (Aqua Products Plc) |
| A Morley | Main Contractor (Leavesden Barnes & Co Ltd) |
| G Mackay | Main Contractor (Leavesden Barnes & Co Ltd) |
| B Hunt | QS (Fussedon, Knowles & Partners) |
| I Tegan | Structural Engineers (GFP & Partners) |
| I Hills | Architect (Reed & Seymore) |
| H A Hemming | Clerk of Works |
| F Adams | Services Engineers (Black and Associates) |

| Item | | Action |
|---|---|---|
| 1.0 | **Apologies**<br>None | |
| 2.0 | **Minutes of last meeting**<br>Agreed as correct | |
| 3.0 | **Contractor Report** | |
| 3.1 | Progress is generally satisfactory | |
| 3.2 | Still one week behind programme due to late delivery of bricks | |
| 3.3 | A1 5 received and actioned | |
| 3.4 | Details of ironmongery revisions required in next two weeks | Arch |
| 4.0 | **Clerk of Works Report** | |
| 4.1 | Concern expressed about poor stacking of bricks | MC |
| | (Continue accordingly through agenda) | |
| 8.0 | **Any other business**<br>None | |
| 9.0 | **Future meetings**<br>Site Meeting - 4 November 1997, 10:00 am on site<br>Architect's Inspection - 21 October 1997, 10:00am | All |

**Distribution:**

| | | | | | |
|---|---|---|---|---|---|
| 2 | Client | 1 | Service Engineer | 1 | Planning Supervisor |
| 3 | Main Contractor | 1 | Clerk of Work | 1 | File |
| 1 | Quantity Surveyor | 1 | Structural Engineer | | |

# Chapter 4
# Site Duties

## Supervision and inspection

It is important to distinguish between these two activities.

Supervision is the responsibility of the main contractor; he is bound by the conditions of his contract to complete the work in a certain time and to a specified standard.

The architect, under the terms of his appointment, is required to visit the site, at intervals appropriate to the stage of construction, to inspect the progress and quality of the works and to determine that they are being executed in accordance with the contract documents. While on site, it is also often possible to make sure that problems, which are bound to arise, even on the smallest jobs, are satisfactorily resolved. However, the architect is not required to make frequent or constant inspections.

## Inspections

The nature and extent of inspection arrangements will depend on the size and complexity of the works. On a small contract, for example, the builder may be able to rely on a competent general foreman, while on larger contracts, if the job is to be properly organised, a number of foremen and assistants may be required working under a site agent or a contracts manager.

Similarly the architect on a small contract may be able to undertake his normal duties by periodic visits, whereas, on large and complex buildings, one or more clerks of works may be needed and possibly a resident architect. In this chapter we are assuming that there will be clerk of works acting, as the contract provides, 'as inspector on behalf of the employer under the directions of the architect'.

The role of the clerk of works may vary slightly depending on whether the appointment is made by the employer or the architect. If appointed

by the employer, the role may approach that of a project manager, looking after his broader interests.

Inspections can be considered under the following headings:

- formal site meetings
- routine site inspections
- consultant inspections
- inspections by statutory officials
- records and reports
- samples and testing

Formal site meetings have been dealt with in the preceding chapter. The remaining headings, however, do call for some further comment.

# Routine site inspections

The architect will normally have more time to inspect the work when making routine inspections than on those occasions when a formal site meeting is held. The frequency of these inspections will depend on the size and complexity of the work and the speed of the progress being made.

As a matter of safety and courtesy, the architect should make his presence on the site known to the clerk of works and the site agent, who will normally accompany him around the job. The architect should never give instructions to a workman direct. The main contractor, through his site agent or general foreman, is the only one with authority under the contract to receive and act upon architect's instructions. All instructions should be confirmed in writing after the visit, as set out in Chapter 5.

The Building Act 1984 consolidates the building control provisions of a number of earlier Acts, which imposed certain legal responsibilities upon the contractor and employer. Where the possession of the site is vested in the contractor the primary liability for health and safety rests with him, and the architect should ensure that these responsibilities are met. The extent of these roles, formally referred to as the planning supervisor and the principal contractor, is codified under the CDM Regulations. Where the site is in the possession of the employer, or jointly with the contractor, both the employer and contractor are responsible for the health and safety of the workforce and other persons on site. The contractor, as the expert, is responsible for the plant. Recommendations contained in BS 5306: Part 0: 1986 (British Standard for fire extinguishing installations and equipment on premises) should also be observed, a point which might have been covered in the bills of quantities.

When making a site inspection it is easy to be distracted and to overlook items which require attention. It is a good plan therefore, to list in advance particular points to be looked at and any special reason for doing so. For this purpose a standard check-list is helpful as it serves as a useful *aide mémoire*. The contents of such a list must be of a rather general nature and it can then be amplified or adapted for each job according to the type of work and form of construction involved. A typical standard check-list is given in *Example 7*.

# Consultant inspections

In addition to the architect, on larger contracts, the structural engineer and services consultants will make regular site inspections. Dependent on the nature of the work, it is often necessary for the structural engineer to make frequent visits at the start of a contract – for example to confirm or adjust foundation designs following excavation or to advise on remedial works upon opening up of works to existing buildings. The structural engineer will be particularly interested in the testing of concrete and other structural elements (see samples and testing below).

Once the services installations are under way, the services consultants are likely to undertake inspections to ensure that the works are meeting the specification.

## Inspections by statutory officials

Visits by statutory officials are to be anticipated on any contract – with a wider range visiting the larger project. Foremost is the building control officer from the local authority. The contractor is required to give him notice of particular critical points within the construction process, whereupon he will visit and advise/comment upon matters pertinent to the building regulations, for example, foundations, drainage, structure, fire protection, etc. It is common for an inspector to issue directions, but the contractor must refer any such direction to the architect for clarification and confirmation as an instruction. The architect may wish to consider the point for several reasons:

- any effect on the programme
- the cost implication
- the possibility of an alternative solution
- whether it is actually necessary or he wishes to discuss the point with the building control officer.

A local authority environmental health officer may wish to inspect the site on the grounds of noise or dust pollution.

An officer from the Health & Safety Inspectorate may visit to examine the safety of site personnel, for example, scaffolding, working conditions, the safety of the site, access, etc. – especially in respect of the CDM regulations.

Subject to the type of building, a Fire and Civil Defence Authority (FCDA) officer may visit to ensure that the means of escape are acceptable, alarms are audible, fire doors adequate, etc.

In addition to statutory officials, it may be that inspections are sought by others, for example, NHBC inspectors, adjoining owner's surveyors, or surveyors acting for the employer's financial backers.

Whoever inspects the site, it is important that the contractor verifies the purpose of their visit and checks whether it is advisable for any other party to be present. The contractor should always record the visit and make a brief report of the substance and outcome of the visit.

## Records and reports

Where a clerk of works is employed, the architect should receive weekly reports from the site, covering:

- number of men employed in the various trades
- state of the weather and particulars of time lost due to adverse weather conditions
- principal deliveries of materials and particulars of any shortages
- plant on the site
- particulars of any drawings or other information required
- visitors to the site
- general progress in relation to the programme
- any other matters affecting the smooth running of the contract

A standard form of report is available from the Institute of Clerks of Works (as shown in *Example 8*), alternatively, one may be provided by the architect or by the contractor.

The weekly report is valuable in keeping the architect fully informed of day-to-day progress, and also constitutes a useful record for reference if disputes arise at a later date. The weekly report should be compiled from the diary of the clerk of works. This diary should be provided by the architect at the commencement of the job and should be returned to him on completion. In it should be recorded daily all matters affecting the contract.

In the absence of a clerk of works, the contractor should be required to compile a similar factual record of conditions and activities on site and the design team should maintain their own records of site observations, including plant, labour, activity and progress, to augment or amplify the contractor's records.

A copy of the builder's programme should be kept in the office of the clerk of works, and every week the actual progress should be checked and recorded against this programme.

It is the job of the clerk of works to keep records of any departures from the production information which may be found necessary in order that, on completion, the architect has all the information necessary to enable him to issue to the employer an accurate set of drawings of the finished building. These records are particularly important where the work is to be concealed, as for example foundations, the depth of which may vary from that shown on the original production information.

Progress photographs of the work also form valuable records if taken regularly. This is probably best arranged in conjunction with the contractor so that both he and the architect can have the benefit of them.

## Samples and testing

The architect may call for samples of various components and materials used in the building to be submitted for approval, in order that he can satisfy himself of their construction or quality and that they meet the requirements of the client and, where applicable, the local authority.

Some of the items, of which samples would normally be required, are:

● External materials, e.g. facing bricks, artificial or natural stone, precast concrete, marble, terrazzo, slates or roofing tiles.
● Internal finishes, e.g. joinery, mouldings, timber, wall or flooring tiles, other decorative finishes
● Services, e.g. plumbing components, sanitary goods, ironmongery, electrical fittings.

In addition to samples of individual components or materials, the architect will often require sample panels prepared on the site to enable him to judge the effect of the materials in the positions in which they will be used. Panels of facing bricks are an example of this, demonstrating a particular bond or pattern, with different coloured mortars.

It is quite normal to require laboratory tests of basic materials such as concrete. The crushing strength of bricks may also be tested in a laboratory, although, unless the design requirements are stringent, a certificate from the manufacturers giving their characteristics may suffice. The testing of concrete should be carried out on a regular basis and cubes should be cast from each main batch of concrete, carefully labelled and identified. The tests themselves should be carried out by an approval laboratory and test reports should be submitted by the contractor to the architect or structural engineer. Full instructions for such testing procedures are usually included in the specification.

British Standard Specifications, Agrément Certificates and Codes of Practice are specified for many building materials, components and processes, and in carrying out tests it is essential to refer to the appropriate standard or code to ensure that the requirements are complied with. In addition the British Standards Institution lays down acceptable tolerances for manufactured goods, and copies of the relevant standards and codes should always be kept by the clerk of works on the site.

# EXAMPLE 7: STANDARD CHECK-LIST FOR SITE INSPECTIONS

This list should be amplified or adapted according to the nature of the project and may serve the architect or engineer. Some of the items should be inspected jointly with other consultants.

*(1) General*

In all cases check that the work complies with the latest drawings and specification, with the latest requirements of the statutory undertakers and with the building regulations. Ensure that all information is complete.

*(2) Preliminary works*

- scaffolding and the Building Act 1984
- siting of workmen's canteens and builder's offices, etc.
- removal of top soil and location of spoil heaps
- perimeter fencing or hoardings
- provision for protection of rights of way
- protection of trees and other special site features
- party wall agreements and protection of adjoining property
- protection of materials on site
- site security generally
- suitability and siting of clerk of works' or site architect's office
- agree bench mark or level pegs
- agree setting out (responsibility of the contractor)

*(3) Demolition*

- extent
- adequacy of shoring
- preservation of certain materials and special items

*(4) Excavation and foundations*

- width of trenches
- depths of excavations
- nature of ground in relation to trial hole report
- stability of excavations
- pumping arrangements
- risk to adjoining property or general public

- quality of concrete and thickness of beds
- suitability of hardcore (freedom from rubbish)
- quality of sand and ballast (freedom from loam and correct grading)
- damp-proof membranes or asphalt tanking
- ducts, drains or services under building
- correct placing of reinforcement, including diameter, bending and spacing of bars
- consolidation of backfilling, particularly suitability of material used for backfilling
- ascertain depths of piles and driving conditions in case of piled foundations

*(5)  Drainage*

- depths of inverts and gradients of falls
- thickness and type of bed and jointing of pipes
- quality of bricks for manholes and rendering thereto
- testing of drains and manholes (water test), etc.

*(6)  Brickwork, blockwork and concrete masonry*

- approve sample panels of facings and fairface work
- quality and colour of mortar and pointing
- test report on crushing strength where necessary
- *BS* certificates on load-bearing blocks
- position and type of wall ties; inspect cavities to external walls to ensure cavities clean above DPCs and wall ties clean
- correct setting-out and maintenance of regular vertical and horizontal joints
- type and quality of DPCs and correct placing
- correct setting-out and fixings for door frames, windows, etc.
- bedding and levels of lintels over openings
- expansion joints

*(7)  In-situ concrete*

- setting out and stability of shuttering
- correct shuttering to achieve type of finish specified
- setting out of reinforcement, fixings, holes and water bars
- mix and correct procedure of taking test cubes
- curing of concrete and striking of shuttering

*(8) Precast concrete*

- size and shape of units
- finish
- position of fixings, holes, etc.
- damage in transit and erection

*(9) Carpentry and joinery*

- freedom from loose knots, shakes, sapwood, insect attack, etc.
- dimensions within permissible tolerances
- application of timber preservatives and primers
- storage and stacking and protection from weather
- jointing, bolting, spiking and notching of carpenter's timber
- spacing of floor and ceiling joists and position of trimmers
- spacing of battens, position of noggings
- weather throatings and hardwood cills to doors and windows, etc.
- jointing, machining and finish of manufactured joinery

*(10) Roofing*

- pitch of roof
- spacing of rafters and tile battens
- approval of underfelt
- approval of roofing materials and fixings
- pointing to verges and bedding of ridges, etc.
- falls to outlets for flat roofs
- thickness and fixing of insulation under finish
- eaves details and ventilation
- correct formation of flashings, etc.

*(11) Cladding*

- vapour barriers and insulation
- regularity of grounds for sheet materials
- location and quality of fixings (cleats, bolts, nails, etc.)
- laps, tolerances, and positions of joints
- setting-out and jointing of mullions and rails
- handling and protection of panel materials (metal, glass and GRP)
- specification and application of mastic
- flashings, edge trims and weather drips
- entry and egress of moisture

- prevention of electrolytic action
- location and type of movement joints (brick and tile claddings)

*(12) Steelwork*

- sizes of steel
- rivets and welding
- position of steels
- plumbing, squaring and levelling of steel frame
- priming and protection

*(13) Metalwork*

- sizing and spacing of members
- galvanising or rustproofing, if specified
- stability of supports, including caulking or plugging
- isolation from corrosive materials, etc.

*(14) Plumbing and sanitary goods*

- inspect sanitary goods to ensure freedom from cracks or deformities
- location, venting and fixing of stack pipes
- falls to waste branches
- use of traps
- jointing of pipes
- smoke and/or water tests
- location and accessibility of valves, stop-cocks
- drain-down cocks at lowest point
- access to traps and rodding eyes

*(15) Heating, hot water and ventilation installations*

- type of boiler, cylinder, tanks, fans, etc, as specified
- types of pipes
- position and type of stop valves
- position of pipe runs and ventilation trunking
- insulation of pipe work
- identification and labelling of pipes, valves, etc. and directions of flow

*(16) Electrical installation*

- approval of components
- switches, fuses, cables, etc., as specified
- position of heating, lighting, communications and switch points as compared with drawings
- earthing of installation
- lightning conductor installation
- runs of cables and quality of connections, etc.
- labelling and identification of switchgear, etc.

*(17) Specialist installations*

- specialist drawings on site
- drawings of builder's work in connection with installation
- power and plant to be provided during installation
- access for equipment and provision of adequate working space
- temporary support, loading on structure, lifting tackle, etc.
- attendance on site and correct sequence of work

*(18) Paving and floor tiling*

- approve materials
- quality of screeds to receive flooring
- junctions of differing floor finishes
- regularity of finish
- falls to gulleys, etc.
- skirtings and coves

*(19) Plastering*

- storage of material
- correct mix
- preparation of surface
- true surfaces and arrises (including angle and casing beads)
- fixing of laths or plasterboard
- filling and scrimming of joints in plasterboard
- bonding plaster on concrete or adequate hacking
- wall tiling, regularity of joints
- external angles

*(20) Suspended ceilings*

- type of suspension and tile
- height of ceiling and correct setting out
- position of light fittings, ventilation grilles, etc.
- access panels
- finish for curtains, blinds, etc.

*(21) Glazing*

- quality of glass and freedom from defects
- correct type and/or thickness
- depth of rebates
- glazing compound, fixing of glazing beads, etc.

*(22) Painting and decorating*

- approval of material being used
- preparation of surfaces and freedom from damp
- inspect surfaces partly concealed, viz insides of eaves, gutters
- finished work – freedom from runs, brush marks, etc.
- opacity of finish, etc.

*(23) Cleaning down and handing over*

- windows cleaned and floors scrubbed
- sanitary goods washed and flushed
- painted surfaces immaculate
- doors correctly fitted, windows not binding or rattling
- ironmongery complete and locks and latches operating correctly
- correct number of keys
- connection of services, provision of meters, etc.
- commissioning of all mechanical engineering plant, balancing of air-conditioning, etc.
- plant maintenance manuals, plant room service diagrams, etc.
- operation of security, communication and fire protection systems

# EXAMPLE 8

## CLERK OF WORKS PROJECT REPORT  NO 6

INSTITUTE OF
CLERKS OF WORKS

PROJECT: Shops & offices .......... REF: ..............

ADDRESS: Newbridge Street Borchester .........

Architect/Contract Administrator Reed & Seymore   Week/~~Month~~ Ending 15 December 1996

Contract Start Date 4 November 1996

Main Contractor Leavesden Barnes   Contract Completion Date 11 September 1997

Clerk of Works Phone/Fax No ..........   Progress +/- to Programme ..........

| TRADES | Mon | Tues | Wed | Thur | Fri | Sat | Sun |
|---|---|---|---|---|---|---|---|
| Site Staff | 3 | 3 | 2 | 3 | 3 | | |
| Groundworkers | 2 | 2 | 2 | 2 | 2 | | |
| Steelfixers | | | | | | | |
| Steel Erectors | 2 | 2 | 2 | | | | |
| Concretors | | | | | | | |
| Drainlayers | 1 | 1 | 1 | 1 | 1 | | |
| Machine Operators | | | | | | | |
| Carpenters | 2 | 2 | 2 | 1 | 2 | | |
| Scaffolders | | | | | | | |
| Bricklayers | 4 | 4 | 4 | 2 | 2 | | |
| Roof Finishers | | | | | | | |
| Wall Cladding | | | | | | | |
| Window Fixers | | | | | | | |
| Glaziers | | | | | | | |
| Floor Screeders | | | | | | | |
| Plasterers | | | | | | | |
| Tilers-Wall/Floor | | | | | | | |
| Dryliners/Partitions | | | | | | | |
| Ceiling Fixers | | | | | | | |
| Decorators | | | | | | | |
| Floor Finishers | | | | | | | |
| Heating/Ventilation | | | | | | | |
| Plumbing | | | | | | | |
| Electricians | | | | | | | |
| Hard/Soft Landscape | | | | | | | |
| Roadworks | | | | | | | |
| Public Services | | | | | | | |
| Site agent | 1 | 1 | | 1 | 1 | | |
| TOTAL | 15 | 15 | 13 | 10 | 11 | | |

Contractors Labour Returns May Be Substituted

**Delays including Defective Work (Action Taken)**
3m length of drainage trench collapsed (NE corner) 11 Dec. Re-excavation required with sheet piling. 1 day lost.

**Site Directions Issued**

| No | Item | Date |
|---|---|---|
| 4 | Manhole 3 repositioned | 9 Dec |

**Drawings/Information Received on Site**
Roof details - dwg nos. 456/55A, 56B, 57

**Drawings/Information Required on Site**
Finishes details

**Plant/Materials Delivered to Site or Removed**
Steelwork delivered

**General Comments**
Steelwork primer badly scratched

### WEATHER REPORT

| | AM | °C | PM | °C | Time Lost |
|---|---|---|---|---|---|
| Mon | Fine | 5 | Fine | 6 | |
| Tues | " | 4 | " | 5 | |
| Wed | overcast | 4 | overcast | 4 | |
| Thur | heavy rain | 4 | rain | 3 | 1 day |
| Fri | Fine | 3 | Fine | 3 | |
| Sat | | | | | |
| Sun | | | | | |

Stoppages (Hours)

Total to Date

| Visitors (Include Statutory Inspectors) | Date |
|---|---|
| Building Inspector | 9 Dec |
| Project Architect | 11 Dec |
| Health & Safety Inspector | 11 Dec |

# EXAMPLE 8 *cont.*

| | Progress to date % | Programme % | | Progress to date % | Programme % | | Progress to date % | Programme % |
|---|---|---|---|---|---|---|---|---|
| Preliminaries | | | Blockwork Internal | | | Electrical 2nd Fix | | |
| Excavation | 80 | 85 | Cladding/Curtain Wall | | | H & V 2nd Fix | | |
| Shutter/Reinf | | | Windows-glazing | | | Ceiling Grid/Tiles | | |
| Concrete Structure | | | Joinery 1st Fix | | | Decoration | | |
| Steel Erection | 15 | 15 | Plastering | | | External Works | | |
| Main Drainage m/h | 50 | 60 | Drylining/Partitions | | | Hard/Soft Landscape | | |
| Floor Construction | | | Floor Screeds | | | Roadworks | | |
| Floors Suspended | | | Plumbing 1st Fix | | | Mains; Gas-Electrical | | |
| Roof Structure | | | Electrical 1st Fix | | | Mains; Water-Telecoms | | |
| Roof Coverings | | | H & V 1st Fix | | | Lifts | | |
| Drainage fw. sw. | 25 | 30 | Wall/Floor Tiles | | | Alarm/Computer Systems | | |
| Brickwork External | | | Plumbing 2nd Fix | | | Defects/Handover | | |
| | | | | | | | | |
| | | | | | | | | |

GENERAL REPORT: Summary of Work Proceeding

Excavations largely completed
Drainage progressing well — but see note under "delays"
Building inspector satisfied with footings
Steelwork started this week

Site Conditions/Cleanliness/Health & Safety: Action Taken

Conditions generally good due to recent fine weather
Reason for collapse of drainage trench being investigated

| Enclosures: G.C. Labour Return [ ]   Site Directions [1]   Reports (State) [ ] | OFFICE ACTION: |
|---|---|

Distribution as Agreed
Client [1]   Architect [1]   Project Manager [·]   Quantity Surveyor [1]   Office [ ]   Others [ ]
Clerk of Works  Harry Hemmings     Date  17 December '96

# Chapter 5
# Instructions, Variations and Cost Control

Provided that the procedure set out in *Pre-Contract Practice* is followed, it should be possible for complete sets of drawings, together with specification notes and nominated sub-contractors' and suppliers' estimates, to be available to the quantity surveyor for the preparation of the tender documents. This in turn will mean that immediately the contract is placed the contractor can be handed all the necessary information to build the project.

## Architect's instructions

The ideal circumstances outlined above are not the norm, however, and even in circumstances where fully finalised information is available for incorporation into the contract documents, it will probably still be necessary, from time to time, for the architect to issue further drawings, details and instructions. These are collectively known as 'architect's instructions'. The conditions of contract set out those matters in connection with which the architect is empowered to issue such instructions, and these are as follows:

| Clause | Description |
|---|---|
| 2.3 & 2.4.1 | Discrepancies in documents |
| 6.1.3 | Compliance with statutory requirements |
| 7 | Levels and setting out the works |
| 8.3 | Opening up work for inspection |
| 8.4 & 8.5 | Architect's powers when work, materials or goods not in accordance with the contract |
| 8.6 | Exclusion from the works of any person employed thereon |
| 13.2 | Variations as defined in clause 13.1 |
| 13.3.1 | Expenditure of provisional sums included in the contract bills |

| 13.3.2 | Expenditure of provisional sums included in a nominated sub-contract |
| 13A.4.1 | Pre-priced variations |
| 17.2 & 17.3 | Making good defects, shrinkage and other faults |
| 21.2.1 | Joint names insurance for adjacent buildings |
| 22D.1 | Insurance for employer's loss of liquidated and ascertained damages |
| 23.2 | Postponement of the execution of any works |
| 34.2 | Action in the event of antiquities being found |
| 35.5.2 | Removal of objection to sub-contractor nomination |
| 35.6 | Issue of NSC/N sub-contractor nomination |
| 35.9 | Action upon receipt of a non-compliance notice issued by a contractor in relation to an instruction to nominate a sub-contractor |
| 35.18.1.1 | Nomination of a substitute sub-contractor in the event of a nominated sub-contractor failing to rectify defects |
| 35.24.6 | Procedure following contractor's application to determine a nominated sub-contractor's employment as a result of the sub-contractor's default; and subsequent re-nomination. |
| 35.24.7 | Re-nomination in the event of a nominated sub-contractor's bankruptcy, etc. or if the contractor is required by the employer to determine the employment of a nominated sub-contractor |
| 35.24.8 | Re-nomination in the event that a nominated sub-contractor determines employment under the sub-contract or where, through no fault on the part of the sub-contractor, works already carried out under the sub-contract require to be redone, and the nominated sub-contractor refuses so to do |
| 36.2 | Nomination of nominated suppliers |
| 42.14 | Integration of performance specified work |

The procedure for the issue of instructions is set out in clause 4.3 and may be summarised as follows:

(1) An instruction issued by the architect must be in writing
(2) An oral instruction is not effective unless it is confirmed by the contractor or architect within seven days. If confirmed by the contractor within the stated period and the architect does not dissent then it is deemed to be an architect's instruction, and
(3) The instruction is effective from the date of the issue of the architect's instruction or the expiration of the seven day period referred to above, but

(4) If neither the architect nor the contractor confirms an oral instruction, but the contractor still carries out the work in question as a matter of goodwill, then the architect can at any time, up to the issue of the final certificate, confirm the oral instructions in writing; these instructions should then be effective from the date of the written confirmation.

*'An instruction issued by the architect must be in writing'*

It is important to note that, if a clerk of works is employed on the project and issues any directions to the contractor, such directions are only effective if they are in respect of matters which the architect is empowered to issue instructions, and if they are confirmed by the architect within two working days, not, it will be noted, within seven days as provided for confirmation of the architect's own verbal instructions. Standard forms are available for such clerk of works' directions.

All instructions from the architect to the contractor should be issued, or confirmed, on a standard form. An example of such a form, shown in *Example 9*, is published by RIBA Publications Ltd, but many other forms, containing the same essential features, have been produced for specific jobs or architectural offices. The use of such sequentially numbered forms has several advantages, and is particularly useful for the monitoring and control of post-contract variations.

It contributes to the smooth running of the contract, if verbal instructions given by the architect during site visits, or directions given by the clerk of works, are recorded in a duplicating site instruction book, which can be signed at the time and subsequently confirmed by an instruction as set out above.

It is essential that instructions should be clear and precise and, where revised drawings are issued, the revision should be specifically referred to. Instructions emanating from consultants should be passed to the architect for confirmation by architect's instructions. Copies of architect's instructions should be distributed as follows:

- contractor
- project manager or employer's representative
- quantity surveyor
- clerk of works
- other consultants

It should be stressed, at the initial site meeting, to all concerned, that no adjustment will be made to the contract sum unless the matter is covered by an architect's instruction in accordance with the terms of contract.

Architect's instructions nominating sub-contractors and suppliers are covered by clauses 35.6 and 36.2 respectively and are dealt with in more detail later in this section. It should be noted that clause 13.1.3 precludes the architect from issuing an instruction nominating a sub-contractor to carry out work which has been measured in the bills of quantities and priced by the contractor as general contractor's work.

If the architect wishes to do this, he may do so only with the contractor's agreement. If the contractor does agree, arrangements for his profit, attendance on the sub-contractor and any other financial matters, should be settled before the nomination is made.

In summary, the architect's instruction will describe the varied work, state any item or items to be omitted, and will record any new or amended drawings which are to be worked to. If the instruction involves the adjustment of prime cost sums, particulars must be given of the estimate which is to be accepted, stating the date of the quotation and the reference number, as well as the total value of the proposed accepted estimate.

# Variations

There is a clear distinction between architect's instructions and variations. Whilst all variations arise as a consequence of architect's instructions, not all architect's instructions are variations.

In considering those architect's instructions which constitute variations, it is useful to consider the definition of 'variation', as defined in tabulated form in clause 13, which is set out below:

13.1    The term 'variation' as used in the conditions means:

13.1    .1  the alteration or modification of the design, quality or quantity of the works as shown upon the contract drawings and described by or referred to in the contract bills, including:

13.1    .1.1  the addition omission or substitution of any work;

13.1    .1.2  the alteration of the kind or standard of any of the materials or goods to be used in the works;

13.1    .1.3  the removal from the site of any work executed or materials or goods brought thereon by the contractor for the purpose of the works other than work materials or goods which are not in accordance with the contract;

13.1    .2  the addition, alteration or omission of any obligations or restrictions imposed by the employer in the contract bills in regard to:

13.1    .2.1  access to the site or use of any specific parts of the site;

13.1    .2.2  limitations of working space;

13.1    .2.3  limitations of working hours;

13.1    .2.4  the execution or completion of the work in any specific order; but excludes:

13.1    .3  nomination of a sub-contractor to supply and fix materials or goods or to execute work of which the measured quantities have been set out and priced by the contractor in the contract bills for supply and fixing or execution by the contractor.

The matters set out in clause 13.1.2, relate to the working conditions imposed by the employer, particulars of which will have been set out in the bills of quantities in accordance with Section A of the Standard Method of Measurement SMM7. It should be noted that, whilst there is a general obligation, under clause 4.1.1, for the contractor to comply forthwith with all architect's instructions, the contractor may make reasonable objection to variations which fall under clause 13.1.2.

Clearly, an architect's instruction which constitutes a variation in respect of the matters referred to in clause 13.1.2 could fundamentally alter the timing of the contract activities and the rates and the prices submitted by the contractor. Where this is the case the contractor would not be obliged to comply immediately with the instruction, but would be required to set out, in writing, his reasons for objecting to the instruction. In such circumstances, there may have to be an agreement between the employer and the contractor as to the impact on the contract sum and duration, prior to the implementation of the instruction. If the parties are unable to reach agreement on the matter then the issue could be referred to arbitration, under clause 41.3.3 which provides for such a dispute to be arbitrated upon immediately.

## Valuing variations

All variations to the contract are to be valued by the quantity surveyor. The contract provides that variations and work carried out by the contractor in expenditure of provisional sums and work for which an approximate quantity is included in the contract bills, shall either be valued in accordance with clause 13.5 or, in the case of pre-priced variations, in accordance with clause 13A. The rules to be observed in valuing variations in accordance with clause 13.5 may be summarised as follows:

- Work which can be properly valued by measurement:
    (a) Work similar to that set out in the contract bills, executed under similar conditions and with no significant change in the total quantity, shall be valued at bill rates. This includes work where the approximate quantity in the contract bills is a reasonably accurate forecast of the quantity of work required.
    (b) Work similar to that set out in the contract bills but not executed under similar conditions or where there is a significant change in quantity, shall be valued on the basis of those in the contract bills but with a fair allowance being made for the differences in conditions and/or quantity. This includes work where the approximate quantity in the contract bills is not a reasonably accurate forecast of the quantity of work required.
    (c) Work not similar to that set out in the contract bills shall be valued at fair rates and prices.
- Omissions shall be valued at the rates and prices contained in the bills.
- When valuing the foregoing, the following must be taken into account:
    (a) Measurement must be in accordance with the method of measurement used for the preparation of the bills of quantities.

(b) Allowance must be made for any percentage or lump sum adjustment in the contract bills.

(c) If appropriate, allowance must be made for any additional or reduced preliminaries.

- Work which cannot properly be valued by measurement shall be valued on a daywork basis (see further details below).

- If any variation substantially changes the conditions under which other work is executed, then that other work shall be revalued as if it was itself a variation.

- If the variation does not involve additional or substituted work or straightforward omissions, or if the valuation cannot reasonably be effected by the application of these valuation rules, then a fair valuation must be made.

*... work of similar nature ...*

When valuing variations, it is important to note the provisions of clause 13.5.5 in that, where a variation causes, in the case of other work, either a change in the working conditions, or a significant change in the quantities or in the conditions under which this other work is carried out, then such changes must also be taken into account in the valuation of the variation.

It is important to note that allowance must be made, in valuing variations, for percentage or lump sum adjustments which have been made in the contract bills, for example for profit additions, and, where appropriate, for adjustment also to be made in respect of the additional or reduced preliminaries requirements.

It is not necessary to take account, when valuing variations, any effect which the variations may have on the regular progress of the work, or any direct loss and/or expense which the contractor may have incurred as a result of the variation and which he is unable to recover through the valuation or any other provision of the contract. These matters are dealt with under clause 26.

The valuation of variations to nominated sub-contract works, including the valuation of work carried out against provisional sums included in the sub-contract, is to be made in accordance with the relevant provisions of the sub-contract. The valuation rules in the sub-contract conditions are similar in principle to those of the main contract referred to above, and it is not proposed to deal with them further here. In this connection, however, JCT 80 does clarify the position where the main contractor tenders successfully for nominated sub-contract works. Any variations in the sub-contract works are then valued in accordance with the terms of the sub-contract, not in accordance with the terms of the main contract.

It is interesting to note that, when the contract rules for valuing variations are brought into play, the quantity surveyor has a unilateral responsibility – the contractor is not involved, apart from being entitled to be present when any measurements are made. Should the contractor not be satisfied with the quantity surveyor's valuation, his only formal recourse is to use the dispute resolution procedures set out in the contract. Practically, of course, the quantity surveyor usually works closely with the contractor's surveyor, so that a dispute does not arise and, hopefully, an agreed final account is eventually produced.

## Dayworks

Works valued on a daywork basis effectively involve a cost-plus method of reimbursement, and this is therefore an attractive means of securing payment for variations, from a contractor's viewpoint.

Architects will frequently find themselves in the situation where variation work is recorded by the contractor on daywork sheets and is presented for signature by the architect, even though it may not be clear to the architect which is the most appropriate basis for valuation of the works. Under these circumstances architects should sign daywork sheets 'For record purposes only'.

For works valued on a daywork basis, the records must be submitted to the architect not later than one week following that in which the work was carried out. The architect or his authorised representative, usually the clerk of works, should within a reasonable period of time check the accuracy of the hours recorded and the materials used and, if satisfied, can sign the sheets as a true record of the time and materials used. The daywork sheets will then be passed to the quantity surveyor, who will determine the method of valuation to be used. It is important to remember that a signed daywork sheet does not automatically entitle a contractor to payment on a daywork basis, as it may well be, for example, that the work can be measured and valued in the normal way.

It is essential that all daywork sheets be serially numbered, state the date when the work was carried out, specify the daily time spent on the work, set against each operative's name, and list the materials and plant used. The architect should check that he has issued the necessary instruction covering the work (a signed daywork sheet does not constitute an instruction) and reference to this instruction should be made on the daywork sheets concerned.

It is helpful if the contractor gives the architect advance warning of his intention of recording anything as daywork, so that the architect can arrange for particular notice to be taken of the resources used.

## Nomination of sub-contractors

A nominated sub-contractor is defined as a sub-contractor whose selection is reserved to the architect.

The procedures required to effect a nomination, together with the paperwork which accompanies them, may seem to be cumbersome and long-winded. In fact the nomination procedures were simplified in 1991, and have been drawn up with a view to eliminating the many problems which had arisen in the past. Prior to the publication of the detailed nomination provisions, nominations were commonly made on the basis of inadequate information, and with little regard for the impact of the nomination on the programme and working arrangements of the general contractor.

Because of the importance of ensuring that the nominated sub-contractor's work can readily be integrated into the work of the main contractor, it is preferable that decisions are taken in good time, often even before main contractor tenders are invited. The preparatory work leading to a nomination ought, therefore, to be dealt with in the pre-contract stage in accordance with the procedures set out below. This is not always feasible however, and the nomination provisions provide for sub-contractors to be appointed, either in the contract bills, or subsequently upon the issue of an instruction. The provisions should operate in the following way:

- The architect completes the invitation to tender NSC/T Part 1.
- At the same time, the architect prepares the employer/sub-contractor warranty NSC/W.
- The architect sends NSC/1 together with relevant tender documents (the numbered tender documents) and a copy of the main contract appendix information and NSC/W to sub-contractors from whom he wishes to receive tenders.
- The sub-contractor submits his tender on NSC/T Part 2 and executes the warranty NSC/W and returns these to the architect.
- The architect arranges for the employer to sign, or signs on his behalf, the chosen sub-contractor's tender.
- The employer executes the NSC/W requirement either under hand or as a deed.
- The architect sends to the proposed sub-contractor a certified copy of the agreement.
- The architect issues his nomination instruction on the applicable form NSC/N and sends this to the contractor together with a copy of NSC/T Parts 1 and 2, the numbered tender documents and the completed NSC/W agreement.
- The contractor agrees with the sub-contractor the matters set out in NSC/T Part 3.
- The contractor and sub-contractor enter into an NSC/A sub-contract agreement.

If the contractor and the proposed sub-contractor are unable to agree on the matters set out in NSC/T Part 3, then the contractor must, within ten days of receipt of the nomination instruction, inform the architect in writing, giving reasons for the inability to reach agreement. The architect must then issue such instructions as may be necessary. If the proposed sub-contractor validly withdraws his offer, which he may do within seven days of written notification of the identity of the main contractor, again the contractor must inform the architect in writing and await the architect's instructions. In such circumstances, the architect must, either make a fresh nomination, or issue an instruction requiring as a variation the omission of the work concerned. In which case, the architect may, with prior agreement, instruct the contractor to carry out the work.

The contractor does, of course, still have the right to make reasonable objection to a nominated sub-contractor and, if he wishes to make such an objection, the contract requires him to do so within seven days of receipt of the nomination instruction. It must always be borne in mind that, in accordance with clause 35, a sub-contractor may only be nominated for work covered by a prime cost sum in the bills of quantities, where the sub-contractor is named in the bills, in an instruction regarding the expenditure of a provisional sum, or, subject to certain qualifications, in a variation order.

If, after nomination, a sub-contractor defaults in his performance of the contract, or if he goes into liquidation, the architect must make a fresh nomination. Clause 35.24 sets out the procedure to be followed in such circumstances.

## Nomination of suppliers

The purpose of the nominated supplier provisions, incorporated into clause 36, is to enable the architect to select, during the course of the contract, specific goods or materials which are not fully defined in the bills of quantities.

A supplier is nominated, or deemed to be nominated, if the supply of materials or goods is covered by a prime cost sum in the bills and the supplier is either named in the bills, or subsequently named by the architect in his instruction regarding the expenditure of the prime cost sum; or where, in an instruction regarding the expenditure of a provisional sum, or in a variation order, the architect specifies materials or goods which can only be purchased from one supplier. In the latter case the materials or goods concerned must be made the subject of a prime cost sum in the architect's instruction.

Clause 36 requires the architect to issue instructions for the purpose of nominating a supplier for any materials or goods covered by a prime cost sum and the clause sets out the manner in which the costs to be set against the prime cost sum are to be ascertained.

Clause 36.4 sets out the conditions of sale which the supplier will be required to accept in his contract of sale with the contractor and if the supplier refuses to accept those conditions the contractor cannot be required to accept the nomination.

The JCT Form of Tender for nominated suppliers is Tender TNS/1 and this substantially reproduces, in Schedule 2, the provisions of clauses 36.3 to 36.5.

# Cost control

Cost control may be defined as the controlling measures necessary to ensure that the authorised maximum cost of the project is not exceeded. It is a continuing process following the cost planning and cost control activities exercised during the design period, as discussed in *Pre-Contract Practice*.

Normally, the authorised maximum cost will be represented by the contract sum. However, it is not uncommon for the sum to be varied during construction. For example, a client, building speculatively, may find that a prospective tenant is prepared to pay more for a particular facility, in which case he will give instructions for the cost and time implications of this to be assessed, and, if satisfactory, can approve an adjustment to the authorised cost.

It is essential for the architect and quantity surveyor to establish at the outset the parameters within which the employer wishes to control construction costs.

Employers will normally have a finite limit on the amount of expenditure available for a project and may insist that expenditure does not exceed that limit. This will often entail seeking economies in construction costs during the latter stages of a project, where variations have exhausted any contingency sum. There are other occasions, however, where construction quality is paramount, and in these instances a lower priority in respect of the authorised maximum cost may well be agreed.

In most construction contracts a contingency sum is included to allow for unforeseen items of expenditure, which invariably arise on complex projects. The extent of the contingency sum will vary according to the nature of the project and the perceived risk of additional expenditure. A refurbishment project, for example, would normally be perceived as representing a higher risk of cost escalation than a new building on a green-field site.

Contingency sums generally amount to a sum in the order of 3 to 5 per cent of the contract, although this can be higher for particularly risky projects.

Generally, any contingency sum should only be used to cover the cost of extra work which could not reasonably have been foreseen at the design stage (e.g. extra work below ground level). It should not be used as a fund to pay for design alterations, except with the prior approval of the employer. In circumstances where no items of unforeseen expenditure arise, it should remain unspent at the end of the contract.

To maintain control of costs, it is necessary for the value of all potential variations to be calculated prior to instructions being issued, in order that their effect may be assessed and, if necessary, action may then be taken to minimise their impact on the contract sum. In this respect, the pre-pricing provisions of clause 13A can be of assistance. The cost control process requires close liaison between architect and quantity surveyor and the other consultants. It demands the quantity surveyor's attendance at all relevant meetings, including site meetings, and the submission to him of all correspondence which may have cost implications. Where time allows, it is a useful discipline for all proposed instructions to be discussed with the quantity surveyor for pricing, prior to their formal issue. This allows consideration to be given to the cost effect of an instruction before the expenditure is committed and, possibly, pre-pricing by the contractor. It is also wise for the architect to look ahead and make early decisions on such matters as expenditure against provisional and prime cost sums and the likelihood of variations in the later stages.

The necessity for strict cost control does not eliminate the need for cost studies in areas of construction which have not been finally detailed. If this is done effectively, the likelihood of having to draw on the contingency sum is lessened. Indeed the contingency provision can even be built up, to cope with unforeseen variations or, if none occurs, a saving can be made for the client against the authorised maximum cost, or the surplus can be used to fund employer approved enhancements to the project.

The evaluation of possible variations and the likely outcome of expenditure to be set against the prime cost and provisional sums, should include the assessment of legitimate claims which might be made by the contractor and all these should be taken into account when forecasting final cost.

In the course of his cost control duties, it will normally be necessary for the quantity surveyor to produce financial reports, on a monthly basis, as forecasts of final expenditure. In these he should not only take into account architect's instructions which have been issued, but also variations which are likely to be issued during the course of the works, which may result in further extras or savings. For example, if the district surveyor or fire officer has indicated that he will be making certain requirements that have not been taken into account in the contract, then the quantity surveyor should include an assessment of the effects of these, pending the formal instructions from the architect. An example of a monthly financial report is given in *Example 10*.

The monthly forecasts of final expenditure should identify the amount of the contingency sum still remaining and consideration should be given to the adequacy of any such amount in the circumstances. In the early stages of the work, most of the contingency sum, if it remains unspent, should still be included, but, as the contract progresses, it can be

(1985) in the English court. This is a complicated area of the law and it would be wise to seek legal advice in the event that a dispute over title becomes likely.

# Completion of the works

It will be necessary to arrange a meeting with all concerned to consider the best method of completing the contract. The employer, his professional advisors and all consultants should be at this meeting.

When a liquidator has been appointed he should be kept informed of decisions which have been made. Similarly, if a bond is coupled with the contract, the bond holder should be kept informed. It would be wise, in the light of recent court decisions, for the employer to check the terms of the bond immediately on becoming aware that the contractor is in financial difficulties. The requirements under the bond, for example as to notice, should be complied with.

There are three options available for completing the works – assuming that the employer wishes to do so and does not choose, following the determination of the contractor's employment, not to complete the works – in which case clause 27.7 would be operated. Assuming completion is intended, the options are:

- reinstatement of the original contractor's employment
- novation – in the form of a '27.5.2.1 Agreement'
- a separate completion contract arranged by the employer

### Reinstatement of the original contractor's employment

Under clause 27.3.3, after the contractor's employment has been automatically determined, the parties can agree to the contractor's employment being reinstated. In these circumstances, the works continue under the original arrangements 'as if nothing had happened'. The contractor is subject to all his original obligations (including completion on time) and the employer maintains all his original rights (including the deduction of liquidated damages).

### Novation – in the form of a '27.5.2.1 Agreement'

Following written notification under clause 27.3.2, the parties enter an interim period under clause 27.5, during which their original obligations are suspended and they are free to make interim arrangements for the work to continue, pending a decision to formalise a new arrangement or terminate the contractor's employment.

Clause 27.5.2.1 makes provision for this new agreement to be a continuation of the original contract, a novation or a conditional novation. This allows sufficient flexibility for the parties to negotiate almost any agreement that suits them and the subsequent contract will be governed by that agreement.

It should be remembered that, in most of these situations, an insolvency practitioner will be acting on behalf of the contractor.

## A separate completion contract arranged by the employer

Following determination of the contractor's employment, whether it be automatic or by the employer's option, the provisions of clause 27.6 become operative, allowing the employer to complete the works as economically and expeditiously as possible – assuming that he wishes to do so and does not decide not to do so under clause 27.7.

To facilitate the completion of the works clause 27.6.1 provides for the employer to employ and pay other persons to carry out and complete the works. Such persons may use all temporary buildings, plant, tools, equipment and materials and purchase all additional materials necessary to complete the works and make good defects. These rights of the employer, when based upon a determination through insolvency, are dependent upon, firstly, a valid right in law to use such buildings, plant, tools etc., and secondly, the owners of the various items being bound by the contract. In the first case, it is often argued that this clause is void against the liquidator, who, as a matter of law, upon his appointment takes into his custody and control all the property of the company. In the second case, many items of plant are hired by the contractor or owned by a subsidiary company which is not a party to the contract. In this case it will be necessary to enter into a new agreement with the owners of the hired equipment.

These factors will be taken into account in putting together the documentation and agreeing the arrangements with the completion contractor. Much will depend on the extent of the work to be completed, as follows:

(1) *If work on site has not started,* consider approaching the second lowest tenderer or, if circumstances have changed, seek revised tenders from all on the original list.

(2) *If the contract was only just started,* then it would be reasonable to go out to tender again inviting contractors to tender for the completion of the works. In this case the quantity surveyor could deal with the tendering documents by preparing a bill of the work already

done, which would form an omission bill, to be read with the original bill and together they would form the bills of quantities of the completion contract.

(3) *If the work is well in hand but far from complete*, it may be best to get competitive tenders for completion, though sometimes it may be advantageous to negotiate a contract with a single contractor.

It is important that the employer takes due care to complete the work in a reasonably economical way, otherwise there could be difficulties with the bond holder or liquidator. Where the completion contract has been negotiated, the employer should be able to show that negotiating the contract was more economical than going out to competitive tender again. The time saved by negotiation may well be significant in this respect.

The quantity surveyor will need to decide whether it is quicker for him to produce a new document for the completion of the works or to use the existing bills with an omission bill for work which has already been carried out.

(4) *If the work is substantially complete or complete apart from making good defects*, what remains to be done may be in the nature of jobbing work, for which a completely different type of builder may be more suitable. Further, it may be better to complete the work under a prime cost contract, having a fixed fee. Consideration could also be given to using an employer's own direct labour organisation or maintenance department, if available.

The quantity surveyor will need to assess the work still to be done and take into consideration sub-contractors' work, before making recommendations as to the most suitable procedure and contract for completion.

## Completion documentation

One item that is essential in any scheme for completion of the work, except where the contractor's employment is reinstated, is a method for dealing with the making good of defects which are bound to arise from the original contract. There are three options available:

- a lump sum premium
- a provisional sum
- measured quantities for known defects, plus a lump sum premium or provisional sum for other items.

## Lump sum premium

In certain cases, the most satisfactory completion document is one on identical terms to the original contract, with the addition of a single fixed premium to cover the new contractor's costs in taking over the partly completed contract, including price fluctuations, the responsibility for making good defects and a credit for the materials on site which belong to the employer. This premium is then paid to the contractor by instalments as the work proceeds.

This system has the advantage of identifying the employer's total commitment at the start and avoids the need for duplicated accounting and therefore saves professional fees. Unfortunately, the system cannot be applied if the new contractor is not able to make a proper assessment of the risks involved – usually the extent of the defects to be remedied.

## Provisional sum

Where the remedial works cannot be defined, a provisional sum has to be included to cover any work which comes to light during the course of the completion contract, which is treated as a variation under the contract and priced according to the normal rules for variations.

This system has the disadvantage of an unknown commitment, but avoids the contractor having to assess the risk.

## Measured quantities for known defects, plus a lump sum premium or provisional sum for other items

Where certain defects are known and, perhaps, involve extensive remedial work, it can be helpful to measure and price the necessary remedial work as part of the completion pricing or negotiating procedure. The remaining defects known, or unknown, can then be covered by a premium or provisional sum.

This is a 'halfway house' solution, which has the advantage of identifying a large element of the remedial works costs and reducing the element left at risk to either the contractor (in the case of a premium) or the employer (in the case of the provisional sum). It may well be an appropriate compromise in an uncertain situation, when time available to reach an agreed basis for completing the contract is limited.

Most of the work in preparing the new contract documents will fall to the quantity surveyor. It will probably not be necessary for the architect to do anything other than to reissue his drawings, as it will be obvious to the new contractor that he is only to do the work which has not already been done. For the purpose of establishing a price, however, the quantity surveyor will need to prepare documents in one of the ways referred to above and this should be done as soon as possible.

## Nominated sub-contractors

Under clause 31 of NSC/C, if the employment of the contractor is determined under clause 27 of the main contract, the employment of the sub-contractor is also determined. When this happens there are bound to be a number of sub-contractors to deal with and they will fall into two groups:

(1) *Where the works have not been started* they will be invited to enter into a similar sub-contract with the new contractor when the new contract is placed.

(2) *Where the works have been partially completed* the quantity surveyor should negotiate a price for completing the work, taking care to ensure that only the work to be completed is measured, assessed and valued, making no allowance for the possibility that the sub-contractor may not have been paid for the work already done.

It will clearly be in the employer's interest for the works to be carried out by the original sub-contractors, but it may be that they will not agree to complete unless they are paid for the work carried out under the original contract. In this case, the quantity surveyor may need to assess the likely cost if the work has to be completed by another sub-contractor, even if this would result in a higher cost than the original quotation.

While clause 35.13.5.2 covers the position regarding direct payment to nominated sub-contractors, it should be noted that this right of the employer will cease if the contractor company has had a petition presented against it for winding up, or the shareholders pass a valid resolution for voluntary liquidation (clause 35.13.5.3.4).

## Bond

The contractor may be required to provide a bond (through a bank or insurance company) for the due completion of the work and this is usually set at 10 per cent of the contract sum. Subject to its terms, such a bond was traditionally thought to be available to be used where the employer is put to additional expense as a result of the contractor's liquidation. However, this is now by no means certain, in view of recent findings by the courts that the insolvency of the contractor does not amount to a breach of contract, as there are procedures dealing with it within the contract. Until the uncertainty is removed, this is obviously an area where legal advice should be sought by, or on behalf of, the employer.

Although the surety may be liable for paying the employer such extra money as may be necessary to complete the job, they have no control over the way the contract is completed. However, the employer should complete the work without undue extravagance and it would be prudent to keep the surety informed as the work proceeds.

## Final account

Clause 27.6 deals with the respective rights, duties and responsibilities of the employer and the contractor, so long as the latter's employment has not been reinstated and continued. While this could mean that the contractor would be entitled to payment, in practice this is seldom the case. The employer is entitled to the additional expenses of completing the works, and to any damages which he has suffered by reason of the determination. It appears that this would include any extra fees and expenses which are incurred, and the costs of delay.

For example, the architect may well have to make additional visits to the site, and additional prints may be required. The quantity surveyor will be involved in a great deal of extra measurement, as well as having to sort out the complex final accounts. The contractor has to cover these expenses properly incurred by the employer and the amount of any direct loss and/or expense caused to the employer by the determination, subject to receiving a credit for what he would have been paid had the contract not been determined. Thus, two final accounts have to be prepared.

(1) The first, or notional, final account will be the amount that the contract would have cost had the first contractor completed in the ordinary way. This should include variations ordered during both the original and the completion contract, all priced at the rates relevant to the original contract.

(2) The second final account will be a normal final account for the completion contract, but will include only those variations ordered during the completion contract. These will be priced by reference to the completion contract, the rates of which may well be different from those in the original contract.

Care should be taken to ensure that if the completion contractor has to make good any defects arising from the original contract, the cost of such works, priced as a premium or treated as variations on the completion contract, does not appear as a variation in the first final account, because they would not have been so included, if the first contractor had completed normally.

Two examples of a statement showing the financial position of the parties at completion are illustrated in the cases below.

In case 1, it has been assumed that the contract was completed without too much trouble and that there is still something to be paid to the contractor in liquidation. It has also been assumed that the extra cost of completing the contract was less than the monies outstanding to the contractor at the time of the liquidation and therefore there is finally a debt due from the employer to the liquidator (see p. 136).

In case 2, it has been assumed that, with the contract only one third complete, the extra cost of completing the work was considerably greater than the money outstanding at the time of the liquidation. Almost certainly, the employer suffers a loss; how much depends on the amount in the pound paid by the liquidator, and any amount contributed by the bond holder, if applicable (see p. 136).

The agreement of the second final account should not create any difficulties because there is a contractor's organisation with which to work. But the first final account will almost certainly have been prepared by the quantity surveyor without the assistance of the first contractor, whose staff, following liquidation, will probably no longer be available. The liquidator should therefore be kept informed of the method used to prepare the account and should be sent a copy on completion for his information.

Clause 27.7 covers the procedure in the event that the employer decides not to complete the works. Should the employer decide after the determination of the employment of the contractor not to complete the works, he is required to notify the contractor within a period of six months and thereafter, within a reasonable time, to send a statement of account to the contractor.

## Retention monies

Clause 30.5 gives rules for the treatment of retention. The employer's interest in the retention is fiduciary as trustee for the contractor and for any nominated sub-contractor. This is to protect a contractor and nominated sub-contractors in the event of the employer's insolvency. However, if a contractor becomes insolvent, it may still be open to the

# CASE 1 – *Resulting in debt from employer to contractor*

|  | £ | £ |
|---|---:|---:|
| Amount of original contract (with company in liquidation) ......... |  | 200,000 |
| Additions (whether ordered with company in liquidation or completion contractor, but priced at rates in original contract) ............ |  | 10,000 |
|  |  | 210,000 |
| Omissions (same rules as for additions) ...................... |  | 7,500 |
| Amount of notional final account if original contractor had completed .. |  | 202,500 |
| Amount of completion contract ............................ | 30,000 |  |
| Additions (for completion contract only and priced at relevant rates in the completion contract) ................................ | 2,500 |  |
|  | 32,500 |  |
| Omissions (ditto) ......................................... | 2,000 |  |
| Amount of final account of completion contract ................ | 30,500 |  |
| Additional professional fees incurred ......................... | 1,500 |  |
| Amount certified and paid to original contractor before liquidation .. | 169,000 | 201,000 |
| Debt payable by employer to liquidator or receiver ............. |  | £   1,500 |

# CASE 2 – *Resulting in debt from contractor to employer*

|  | £ | £ |
|---|---:|---:|
| Amount of original contract (with company in liquidation) ......... |  | 200,000 |
| Additions (whether ordered with company in liquidation or completion contractor, but priced at rates in original contract) ............ |  | 10,000 |
|  |  | 210,000 |
| Omissions (same rules as for additions) ...................... |  | 7,500 |
| Amount of notional final account if original contractor had completed .. |  | 202,500 |
| Amount of completion contract ............................ | 150,000 |  |
| Additions (for completion contract only and priced at relevant rates in the completion contract) ................................ | 10,500 |  |
|  | 160.500 |  |
| Omissions (ditto) ......................................... | 7,700 |  |
| Amount of final account of completion contract ................ | 152,800 |  |
| Additional professional fees incurred ......................... | 7,500 |  |
| Amount certified and paid to original contractor before liquidation | 72,500 | 232,800 |
| Debt payable by liquidator or receiver to employer and partly by bond holder (if applicable) ................................... |  | £ 30,300 |

liquidator of the contractor, in a situation where the employer withholds the whole of the retention monies from the contractor, to proceed against the employer as trustee for the nominated sub-contractors. Thus, the employer may not be entitled to have recourse to the retention money held for the benefit of the nominated sub-contractors.

Clause 30.5.3 provides, should the contractor or nominated sub-contractors so request, for the employer to place retention monies in a separate and appropriately designated bank account.

## The procedure upon the insolvency of a nominated sub-contractor

Clause 35.24.7 states that where a nominated sub-contractor becomes insolvent and his employment is determined, the architect shall make such further nomination of a sub-contractor, in accordance with clause 35, as may be necessary to supply and fix materials or goods or to execute the work and to make good or to re-supply or re-execute any defective materials or defective work. See *Bickerton* v *NW Metropolitan Regional Hospital Board* (1970) (which was a decision on the terms of a JCT 1963 contract).

Arising from the decision of the House of Lords in the case of *Percy Bilton* v *GLC* (1982) (again a decision on the terms of a JCT 1963 contract), it would seem reasonable to assume that delay inevitably occurring due to the departure (or dropping out) of a nominated sub-contractor would not entitle the main contractor to any extension of time under clause 25.4.7, but clause 35.24.10 requires the architect to make the further nomination of a sub-contractor within a reasonable time of the obligation having arisen, having regard to all the circumstances.

Insofar as any period of delay could be attributable to failure to nominate in a reasonable time, it would appear that this would now be dealt with under clause 25.4.5. Therefore, an architect has power to extend the contractor's time for completion if a nominated sub-contractor goes into liquidation and there is subsequent undue delay in re-nomination.

In considering what is meant by the term 'within a reasonable time', it is necessary to study the judgement of both the judge at first instance and in the subsequent Court of Appeal in the important case of *Rhuddlan Borough Council* v *Fairclough Building Limited* (1985).

The Court of Appeal dealt with a number of important issues and decided, amongst other matters:

- that a main contractor can refuse to accept the re-nomination of a substitute sub-contractor who does not offer to complete his part of the work within the overall period for the contract

- following the Bickerton case, referred to above, the main contractor is not bound to do any of the sub-contract work himself, unless otherwise agreed. This means that the employer must order its omission or issue a variation order and pay the contractor for it, or negotiate a new sub-contract covering remedial and uncompleted work

## The procedure upon the insolvency of the employer

Clause 28.3 deals with the insolvency of the employer in a manner similar to the way that clause 27.3 deals with the insolvency of the contractor.

Clause 28.3.2 requires the employer to inform the contractor in writing if he makes a composition or enters into a voluntary arrangement.

Clause 28.3.3 provides for the contractor to give notice to the employer determining his employment following any act of insolvency, bearing in mind that, notwithstanding any formal notices following any of the events set out in clause 28.3.1, the obligation on the contractor to carry out and complete the works in compliance with clause 2.1 is suspended.

Clause 28.4 covers the consequences arising from determination for any reason, including the employer's insolvency. In essence, the contractor is required to remove all his temporary buildings, plant, tools, equipment and site materials safely, and ensure that sub-contractors do so.

As far as the works are concerned, they are unlikely to proceed, and clause 28.4.3 requires the contractor to prepare an account for submission to the employer and, after taking into account amounts previously paid under the contract, the employer shall pay to the contractor the amount properly due within 28 days of submission.

The employment of the architect and the quantity surveyor will not necessarily end with the employer's liquidation – their terms of appointment should be checked. However, if they are to continue, they should seek undertakings from the liquidator that they will be paid for their services. If the liquidator is prepared to give such undertakings they can then proceed with their respective duties in the winding down of the contract, the settlement of accounts and resolution of claims.

# Index